PEOPLE, PLANTS, and LANDSCAPES

PEOPLE, PLANTS, and LANDSCAPES

Studies in Paleoethnobotany

Edited by
Kristen J. Gremillion

The University of Alabama Press
Tuscaloosa

∞

The illustration on the cover is a deerskin map with Charlestown
(Charleston), South Carolina. (English copy, c. 1721. Colonial
Office Library 700, North American Colonies, General No. 6(1),
Public Records Office, Kew, London. Photographic duplicate pro-
vided by the North Carolina Collection, University of North
Carolina Library at Chapel Hill.)

The paper on which this book is printed meets the minimum
requirements of American National Standard for Information
Science–Permanence of Paper for Printed Library Materials,
ANSI Z39.48-1984.

Library of Congress Cataloging-in-Publication Data

People, plants, and landscapes : studies in paleoethnobotany / edited
by Kristen J. Gremillion
p.   cm.
Includes bibliographical references and index.
ISBN 0–8173–0827–X (paper : alk. paper)
1. Paleoethnobotany.   I. Gremillion, Kristen J., 1958–
CC79.5.P5P45   1997
581.6—dc20   96-13617

To Dick Yarnell

with affection and gratitude

CONTENTS

Contents

FIGURES AND TABLES

Figures

Figures and Tables

Tables

FOREWORD

BRUCE D. SMITH

Science, and our understanding of past human societies, does not advance gradually and uniformly across a broad front of inquiry. Rather the advance occurs as rapid and exciting expansions in some areas along the front while in other areas nothing much may happen for long periods of time. Researchers who move into these areas of rapid advance, in which broad new fields of inquiry are opening up, are often faced with exciting and challenging opportunities to apply emerging new theory and technology to expanding and complex data sets.

This book brings together a rich diversity of different case studies carried out within one such region of productive research that has steadily expanded over the past thirty years: archaeobotany. Plant remains recovered from archaeological contexts have long held the promise of illuminating a variety of different aspects of past human societies, but it is only in the past thirty years that water flotation, accelerator mass spectrometry dating, scanning electron microscope analysis, and a range of other innovations have fueled the continuing rapid expansion of this region of inquiry. As Patty Jo Watson discusses in the opening chapter of this volume, many of the advances in archaeobotany over the past three decades have taken place in the Near East and in eastern North America, and in both areas a few key individuals have played central roles.

In eastern North America, Richard A. Yarnell has been a central figure throughout this period. In a variety of ways, both direct and indirect, he has charted the course of archaeobotanical research for the region and for a number of generations of scholars. In this regard the chapters comprising this book provide eloquent testimony to the enduring influence and importance of Richard Yarnell's approach, ideas, and interpretive perspective. There is much in this book, I would guess, that will please Yarnell far more than any glowing praise for him. There are ample and provocative theories and interpretations to be discussed and dissected, careful and well-reasoned arguments to consider, and a wealth of new data across a broad temporal and geographical span: in short, all the ingredients of studies that will stand the test of time.

PREFACE

KRISTEN J. GREMILLION

Origin and Scope

This book had its genesis in a symposium held to honor the recipient of the Fryxell Award for Interdisciplinary Research during the Annual Meeting of the Society for American Archaeology in Pittsburgh in 1992. In that year the award went to Richard A. Yarnell of the University of North Carolina at Chapel Hill. A graduate of the Department of Anthropology at the University of Michigan, Yarnell played a central role in the professionalization of the specialty of paleoethnobotany in the eastern United States. Although he never regarded himself as an archaeologist, Yarnell has pursued his interests in ethnobotany and human ecology primarily by way of analysis of archaeological plant remains and interpretation of the resulting data. His encyclopedic knowledge of the archaeobotanical record of the Eastern Woodlands is legendary, and archaeologists of various stripes continue to consult him on the finer points of cultigen chronology and the status of the latest addition to the roster of prehistoric domesticates in the region. The root (and result) of this notoriety is that Yarnell always knows the biggest, the earliest, the farthest east, west, north, or south, and the most abundant. Because of his status in the profession and his lively and continuing interest in the growing paleoethnobotanical data base, researchers all over the United States send him results and copies of unpublished reports as well as questions, making his office sort of a central clearinghouse of current archaeobotanical research. In addition, his gentle skepticism, persistent common sense, and empirical outlook have helped to keep many a student tethered to reality. Not that he is any stranger to the creative element in anthropological theorizing; the intricacy and logic of Yarnell's views on ecological and evolutionary processes have through the years stimulated and fascinated his students.

Two themes that have been central to Yarnell's research, human modification of the environment and the evolution of plant husbandry, provided the keynotes for the 1992 Fryxell Symposium, which was titled "Cultural and Ecological Dimensions of Anthropogenic Environments and Plant Domestication." Participants, most of whom were students or close colleagues of Yarnell, were invited to apply their own perspectives and technical skills to the symposium's topics. Because of the interdisciplinary theme of the symposium and the shared influence of Yarnell's work, the papers produced brought a variety of lines of evidence to bear on related questions and in many cases represented the integration of data and methods from different disciplines (or subdisciplines within anthropology). The result was a methodologically diverse but thematically coherent set of papers that showcased the synergistic potential of modern paleoethnobotany.

The desire to illustrate that potential for the community of interested scholars led to plans for an edited volume whose core is composed of papers presented in the symposium. In its various chapters, cross-fertilization between areas of study is reflected in two important ways. The first is related to the fact that paleoethnobotany (usually defined as the study of interactions between humans and plants using archaeological evidence) is inherently interdisciplinary (Pearsall 1989:ix). Thus it combines a single area of investigation (human-plant interaction) with multiple methodological and theoretical approaches (including archaeobotany, the recovery and identification of plant remains from archaeological sites [Popper and Hastorf 1988:2]). The majority of the papers reflect methodological and theoretical hybridization by applying botanical knowledge to the study of archaeological plant remains and interpretation of the data they yield, whereas others do so by joining the central subject of paleoethnobotany with nonarchaeological sources of evidence. In addition, many employ techniques (such as accelerator radiocarbon dating) whose sources lie beyond the boundaries of either archaeology or botany. Second, the technically and methodologically enhanced paleoethnobotany of the 1990s has joined forces with ecological and evolutionary theory to forge explanations of changing relationships between human and plant populations. This union is represented by the application of evolutionary ecology approaches to the origins of food production and the selection of plant resources; the use of the concepts of adaptation and anthropogenesis to explore changes in land use and plant management; and the combination of archaeobotanical data and the theory of natural selection to document and explain morphological change in populations of evolving domesticates.

As the book project evolved, some changes occurred in the roster of contributors. All symposium participants were invited to submit written versions of their presentations. In addition, Gayle Fritz (who organized the symposium) developed a chapter. One original participant, Jeff Chapman, was unfortunately unable to contribute to the book, and Bruce Smith agreed to write a Foreword in lieu of the paper he presented in the symposium. The addition of a chapter by Wes Cowan proved to be an unanticipated bonus.

Although the chapters in it reflect intradisciplinary and interdisciplinary cooperation and communication, this book is exclusively the work of anthropologists. Thus despite the incorporation of concepts, methods, and data from botany, genetics, ecology, and history (among others), the contributions are united in their implicit understanding that human behavior is patterned and linked in complex ways to the transmission of information by nongenetic means. They also share the assumption that, the perils of subjectivity and cultural bias notwithstanding, the causes of human behavior are ultimately knowable and can be investigated empirically. The fact that most (though not all) of the authors are archaeologists also contributes to the strong emphasis upon the material conditions of human existence and their centrality to cultural change. Of course, the reasons for behavioral variation are much contested even (perhaps especially) within anthropology. That diversity is reflected in the wide range of explanatory frameworks represented in this book.

Many of the chapters in this volume discuss archaeobotanical evidence for the origins and development of premaize agricultural systems in eastern North America. However, in others at least one of these three recurrent topical ingredients (archaeobotany, native agriculture, and the East) is replaced. Thus the central issue of plant domestication and utilization by past human groups is also explored using nonarchaeological lines of evidence and with reference to other cultural contexts, time periods, and geographical settings. For this reason, this book is potentially of interest to all researchers concerned with the causes and consequences of agricultural origins and human subsistence behavior in ecological context.

Acknowledgments

This volume could not have come together in its present form without the organization of the Fryxell Symposium as a tribute to the career of Dick Yarnell. The Fryxell Award committee of the Society for American Archaeology, 1992, is to be commended for its decision. Gayle Fritz took on the task of organizing the symposium, and this volume owes its origin to her hard work in soliciting papers. I thank all the symposium participants for their excellence and especially those who agreed to contribute to the published version and saw the project through to completion. I am grateful to symposium nonparticipant Wes Cowan for offering a manuscript for inclusion. Judith Knight of the University of Alabama Press offered much useful advice even before the project was fully under way. The comments of Lee Newsom and one anonymous reviewer provided direction for revisions that substantially improved the quality of the finished product.

Last, and far from least, I want to salute Dick Yarnell for accomplishing something that many of our colleagues find elusive: making a significant contribution to scholarship without sacrificing good fellowship, the joy of discovery, or the sheer fun of living. In that, may he serve as a role model for all of us.

PEOPLE, PLANTS, and LANDSCAPES

Introduction

KRISTEN J. GREMILLION

The Development and Research
Potential of Paleoethnobotany

When Richard Yarnell's dissertation was published as *Aboriginal Relationships Between Culture and Plant Life in the Upper Great Lakes Region* in the University of Michigan Museum of Anthropology's Anthropological Papers Series in 1964, archaeological evidence for prehistoric plant use was still relatively limited in quality and quantity. Although detailed ethnobotanical information was available for some areas in and near the Eastern Woodlands (e.g., Densmore 1928; Gilmore 1977; Wilson 1987; see also references in Yarnell 1964), these studies were largely descriptive and synchronic. In the absence of written records, changes in patterns of plant use could not be effectively described or explained for prehistoric periods without the aid of large collections of archaeological plant remains. Widespread acknowledgment of the relevance of subsistence remains to important questions of cultural change and their routine, systematic collection using specialized techniques such as flotation (Struever 1968) still lay a decade or more in the future when Yarnell received his doctoral diploma (see Watson, this volume). Thus, despite its role in bringing together anthropological and botanical knowledge, the research potential of paleoethnobotany was at that time realized in only a limited way.

Partly because of the inadequacy of the paleoethnobotanical data base, agriculture in the Eastern Woodlands was generally understood to have been relatively late and Mesoamerican-inspired. Although tantalizing glimpses of an indigenous "premaize" agricultural tradition had been available at least since Jones (1936) and Gilmore (1931) published their accounts of and speculations about the botanical remains from dry rockshelters in the rugged uplands of the Ozarks and eastern Kentucky, most remained unconvinced of the cultural significance of this early evidence of farming (Gremillion 1993b).

Crucial support for the independence and economic importance of this indigenous agricultural tradition gradually built during subsequent decades as techniques for recovering, dating, observing, measuring, and classifying archaeological plant remains (many of them involving interdisciplinary collaboration beyond the traditional botany-archaeology axis) were developed and refined. Pivotal roles in this process were played by Yarnell's documentation of changing seed size in sunflower and sumpweed (Yarnell 1972, 1978) and his analyses of archaeobotanical assemblages from Salts Cave (Yarnell 1969, 1974a, 1974b). However, it was not until the 1980s that the long-standing hypothesis of an "Eastern Agricultural Complex" acquired firm empirical support and widespread acceptance from new research and techniques in the areas of dating, microscopy, phytotaxonomy, and morphometrics (Gremillion 1993b; Yarnell 1994). The idea that food production originated in one or a very few locations has gradually been replaced by recognition of the likelihood of parallel evolution of the human-plant relationships that result in domestication (Rindos 1984). Similar trends in the explanation of agricultural origins are evident in treatments of many world regions (Cowan and Watson 1992; Gebauer and Price 1992; Harris and Hillman 1989).

The uses to which paleoethnobotanical data were put in explaining past subsistence behavior were initially limited, not only by a dearth of data, but also by the degree and character of interest in ecological problems in North American anthropology. In the 1930s, when Jones and Gilmore were proposing an indigenous agricultural tradition based on their rockshelter finds (Gilmore 1931; Jones 1936), paleoethnobotanical research had a decidedly descriptive bias. Interpretation was geared toward the reconstruction of past agricultural systems and the timing of their development or introduction rather than employment of ecological concepts to explain the archaeological record of plant use. This orientation remained the norm, at least in the Eastern Woodlands, until the 1950s, when Julian Steward began to make significant inroads into the understanding of relationships between culture and environment. His approach, cultural ecology (Steward 1955), was critical in stimulating archaeologists to seek explanation of parallel events, such as agricultural origins, in similar ecological circumstances. In the biological sciences, ecosystems ecology was enjoying the peak of its vogue during the 1960s, and many of the principles that were proving useful to ecologists (see for example Odum 1971) were soon appropriated by anthropologists eager to explore human-environment interactions. Freely borrowing concepts such as succession, stability, and diversity, archae-ologists in particular became increasingly enamored of the homeostatic models developed in ecology, which seemed to offer a broadly appli-cable key to explaining past subsistence change. The work of Flannery (1965, 1969, 1971) was especially beneficial in leading archaeologists to a recognition of the complexity of relationships between environ-mental and cultural factors and in isolating critical causal components of the transition to agriculture.

Meanwhile, the resurgence of interest in evolutionary theory in anthropology was embodied in the highly influential work of Leslie White (1949, 1959), whose vision of transformational and progressive culture change by means of ever-increasing effectiveness of energy cap-ture and control by human societies owes much to the work of Herbert Spencer and other nineteenth-century social theorists (Dunnell 1980). Cultural evolutionism was further developed in the work of Michigan-ites Sahlins and Service (1960). The influence of their approach was felt in the emerging processual archaeology of the 1960s and 1970s, which incorporated ecological theories and concepts into explanations that viewed prehistoric culture change as a product of adaptation to envi-ronmental conditions. However, answers to the "why" questions of sub-sistence change proved frustratingly elusive to archaeologists, in part because they tended to locate causality in the transformation of sys-tems rather than in the differential persistence of behavioral variability (Dunnell 1980; see also Winterhalder and Goland, this volume). The ecosystems approach presented a similar intellectual roadblock; self-regulating processes, while useful for understanding system function, had little power to explain the kind of directional change that was frequently the subject of archaeological inquiry.

Human evolutionary ecology seemed to offer a way out of this dilemma. Evolutionary ecology approaches change from the standpoint of individual decision making and its consequences for fitness; thus it does not require faith in the existence of some system-level tendency to strive for and maintain an adaptive equilibrium (Smith and Winter-halder 1992). For example, optimal foraging theory includes several general models that predict how variables such as diet breadth and for-aging location will vary under different environmental circumstances, given certain constraints and assumptions. Hypotheses derived from these general models are then tested against ethnographic or archae-ological data, and results are used to refine understanding of the relationships between key variables. Quantitative optimization models have been applied, so far primarily in an exploratory way, to problems of prehistoric subsistence (Gardner 1992; Keene 1979, 1981; Reidhead 1976; Yesner 1981), and the role of archaeobotanical evidence with respect to model results is not always clearly defined and understood. However, because of its systematic methodology for generating predictions and its emphasis on behavioral variability in environmental context, evolutionary ecology holds considerable promise as a heuristic tool for paleoethnobotanists.

Chapter Overview

The related themes of the substantive and theoretical maturation of paleoethnobotany as a multidisciplinary catalyst and the integration of paleoethnobotany with ecological and evolutionary theory provide the focal point of this volume. The presentations of original research in Part I, "The Archaeological Record of Plant Domestication and Utilization," address the development of food production outside of the primary "centers" of domestication as defined by Harlan (1975) and others. These chapters demonstrate how the investigation of indigenous systems of plant husbandry has stimulated the implementation of improved methods of dating, measuring, and observing plant remains whose origins lie for the most part outside of anthropology. They also illustrate the application of the concept of natural selection to the study of morphological change in domesticated plant populations. Part II, "Plant Resources, Human Communities, and Anthropogenic Landscapes," considers plant domestication and management in ecological and cultural context. Although diverse in methodology and theoretical orientation, the chapters in this section all propose explanations of human behavior patterns (such as resource selection, landscape modification and use, and social relations) that are in some way influenced by interactions with plant communities.

Initiating Part I, Patty Jo Watson reviews the history of paleoethnobotany in three regions, the Near East, the southwestern United States, and the Eastern Woodlands. Her survey describes the elevation of this discipline from poor cousin of archaeology to a methodologically and technically well-equipped specialty whose practitioners are often consulted (if less often believed). Although improvements in paleoethnobotanical method and technique are evident worldwide, differences among the three regions in the pace and character of progress in the development of the discipline highlight the significant role of mentoring by senior scholars and the determined pursuit of subsistence-oriented research questions.

Gremillion's chapter continues the exploration of paleoethnobotany's development by placing in historical perspective some sixty years of investigations of archaeobotanical collections from the Newt Kash shelter in eastern Kentucky. Volney Jones's landmark study of materials from this site (Jones 1936) provided an important initial spark to the hypothesis of indigenous plant husbandry, and Gremillion's reanalysis illustrates how application of scanning electron microscopy, systematic sorting, and accelerator dating has enhanced the already considerable research potential of dry rockshelter sites. Through such reevaluations, archaeobotanical collections once abandoned as virtually valueless have acquired a new life as subjects of scientific investigation.

Fritz's chapter also illustrates the potency of the application of new methods (such as radiocarbon dating and scanning electron microscopy) to old paleoethnobotanical collections, in this case from the dry bluff shelters of the Ozark uplands, also widely known among archaeologists for their evidence of early farming. Marble Bluff has produced evidence of Terminal Archaic (late fourth millennium B.P.) plant husbandry in an upland setting. At that site, domesticated sunflower, cucurbits, and sumpweed were recovered in contexts suggesting that storage of selected seed stock was practiced with increasing frequency and competence. The early appearance of domesticates at Marble Bluff exhibits striking parallels with contemporaneous sites on the Cumberland Plateau of eastern Kentucky (Cowan, this volume; Gremillion 1993c and this volume).

Like Gremillion and Fritz, Cowan employs data derived from desiccated plant remains. He uses data on rind and seed morphology of *Cucurbita pepo* to demonstrate prehistoric change in the utilization of this taxon. In eastern Kentucky, a thick-walled domesticate (probably a container crop) evolved as early as ca. 3000 B.P. from a weedy, thin-shelled camp-follower gourd that was probably utilized primarily for its edible seeds. Shortly thereafter, several varieties of *Cucurbita* are identifiable in the archaeological record. After ca. 1500 B.P., a growing preference for fleshy-fruited forms is reflected in an overall reduction in rind thickness. Thus a rather clear set of trends is evident in the use of *Cucurbita* first as a seed source, later as a container crop, and finally as a producer of edible fruits. Cowan's chapter illustrates how consideration of the morphological and ecological correlates of the domestication process can lead to inferences about past behavior that would otherwise be difficult or impossible to make.

Crawford's chapter makes similar use of morphometric data in his discussion of archaeobotanical evidence for plant husbandry and environmental change in northeastern Japan. There, changes in the representation of various weed and crop taxa reflect the development of food production in the context of anthropogenic environmental disturbance over several millennia. In many respects, his work parallels that of Gremillion, Cowan, Fritz, and others who have documented processes of initial domestication on the opposite side of the globe. In large part, it is the shared understanding of the operation of ecological processes such as anthropogenesis that makes interregional comparisons profitable, leading to generalizations about the causes and consequences of food production worldwide.

In its emphasis on the processes of interaction between human and plant populations, Crawford's chapter anticipates a major theme of the chapters that follow it in Part II. The first two of these represent contrasting but complementary styles of conjoining ecological theory and paleoethnobotanical data. Scarry and Steponaitis review the development of the much-studied Moundville polity and its role in landscape modification. Spatial variation in the distribution of maize and other food remains suggests provisioning of the elite in the context of a dual system of production (for household and chief). Risk reduction is thought to have been an important motivation for the establishment of communal fields scattered over the landscape.

The same strategy of field scattering is central to Winterhalder and Goland's chapter, which compares theoretical expectations about the development of agricultural subsistence with data from the Eastern Woodlands. They utilize the diet choice model (a product of optimal foraging theory) to explore the relationships between profitability and density of plant resources and the origins of food production. They propose that, once domesticates enter the diet, increases in their density are accompanied by more frequent use. If these abundant domesticates are low in profitability, population growth will cause depletion of higher-ranked resources and expansion of diet breadth; if they are profitable to exploit, less valuable resources will be dropped. Although the implications for diet breadth differ, in both cases specialization results in increased susceptibility of agricultural populations to risk. These conditions favor a shift in risk-reduction strategies from resource pooling to field dispersion in the context of community-level management. In its emphasis on responses to subsistence risk and uncertainty, this formulation of expectations based on foraging theory resembles risk-minimization models of the economic organization of Mississippian society that have been reached independently from rather different theoretical and empirical starting points (see for example Muller and Stephens 1991; Scarry and Steponaitis, this volume).

Gardner's chapter joins the microeconomic view of foraging behavior that is characteristic of evolutionary ecology with quantified environmental data to predict the role of hickory nuts and acorns under changing conditions of climate and anthropogenic disturbance. He demonstrates how the yield and nutritional characteristics of different mast species are likely to have influenced the organization of exploitation of these important resources. Constraints on mast exploitation such as periodicity of yield, costs of processing, and competition may also have acted to stimulate behavioral responses in the form of storage, social alliances, and dispersal. Gardner's contribution is a synthesis in two respects. First, it illustrates how the models and concepts of evolutionary ecology can be comfortably integrated with archaeological and environmental data to stimulate new research directions and generate testable hypotheses. Second, it simultaneously addresses the questions of what preagricultural foragers were doing for a living during the eastern Archaic and how their interactions with the environment may have been causally related to the origins of food production. In linking these two issues, Gardner's study touches on an area of investigation that is of central importance to the understanding of agricultural origins but is often neglected.

The two final contributions illustrate the benefits of exploiting nonarchaeobotanical sources of evidence in order to document and explain patterns of human-plant interaction. Waselkov uses historical documents in conjunction with archaeological data to trace changes in the environmental setting, size, and socioeconomic role of native agricultural fields in the historic-period Southeast. In the lands occupied by the Creeks and Seminoles, communal fields in bottomlands complemented the smaller household garden plots situated in or near settlements. During the early nineteenth century, as the deerskin market declined, livestock husbandry became increasingly common in native communities as a functional substitute for deer hunting. As cattle-raising proliferated, large communal fields were gradually replaced by isolated upland farmsteads strategically located near grazing range. This case study clearly illustrates the causal link between ecological relationships and patterns of social interaction (in this case, the decline of matrilineage power and its replacement by household autonomy).

Hammett takes a broad geographical perspective, using historical records, information on plant population ecology and phytogeography, and archaeobotanical data to document continentwide trends in plant use. Interregional patterns reflect similar landscape structure (divisible into patches, corridors, and matrices), landscape dynamics (including the use of fire as a management tool), a limited set of economically important plant families, and the nearly simultaneous intensification of maize agriculture in eastern and western North America. This geographically broad perspective on plant use encourages the examination of environmental factors (such as climatic events) that might have played similar causal roles continentwide.

This collection of studies reflects to a great extent the implications of the growing paleoethnobotanical data base, and the study of human-plant interaction in general, for theories of human behavior. New methods of observing, dating, and analyzing archaeological plant remains have prompted the recognition of eastern North America as an independent center of plant domestication. Simultaneously, support for multiple "core areas" for initial food production continues to be produced by archaeological investigations worldwide. These studies have been accompanied by the reevaluation of causal factors operating in the shift to food production as anthropologists become increasingly adept at employing ecological concepts and theories to problems of past human-plant interaction. Most recently, they have begun to explore the use of theories that signal a return to Darwinian, selection-oriented models of cultural change in environmental context. Modern studies of human-plant interaction are enhanced by the use of paleoethnobotanical data to evaluate general models of subsistence change, the consideration of that data base in the light of ecological and evolutionary theory, receptiveness to advances in archaeometric techniques, and the use of documentary and botanical evidence. The continued fusion of these approaches is likely to be a primary strength of paleoethnobotany in the years to come.

Introduction

A Note on Terminology

There are several terms whose usages remain somewhat inconsistent in paleoethnobotanical literature and require explanation. Some ambiguity in classifying plants is expectable, given the recognition of the wild/domesticated dyad as a continuum rather than a dichotomy. However, wherever possible I have tried to restrict the use of the terms *domesticate* and *cultigen* to refer to a plant whose morphologic and genetic makeup has been modified away from the typical wild state as a result of its interaction with humans. In contrast, *crop* and *cultivated plant* are used to refer to any species or population that is planted, harvested, and otherwise managed, regardless of the existence of any morphological indicators of this relationship.

Terms used for the production systems and behaviors involved in plant management have similarly eluded standardization. Here, *agriculture, plant husbandry,* and *farming* refer to any complex of behaviors geared toward the management and production of plant resources; *food production* pertains to those same kinds of activities, but without specifying whether plant or animal resources are involved. The term *horticulture* is sometimes used to refer to small-scale gardening as opposed to field agriculture. *Cultivation* in the narrow sense is reserved for breaking up the soil, but by extension refers to practices involved in encouraging the growth of crops.

Vernacular names for particular plant species are generally less variable, with the exception of the cucurbits (members of the Gourd family). Recent taxonomic revision of *Cucurbita pepo* and refinement of its evolutionary history have greatly complicated cucurbit terminology. In this volume, the following conventions are used: *bottle gourd* refers to the species *Lagenaria siceraria; Cucurbita gourd* or *C. pepo gourd* refers to members of the species *Cucurbita pepo* that have hard, lignified fruit walls; *squash* is reserved for fleshy culinary varieties of *C. pepo;* and *gourd/squash* encompasses *C. pepo* material that cannot be placed into either the hard-walled or fleshy categories. Most authors use these terms with reference to *C. pepo* ssp. *ovifera,* which is thought to be indigenous to the Eastern Woodlands (Decker 1988), and its ancestral lineage. Throughout the volume, scientific names are given in each chapter at the first mention of a taxon.

PART 1

The Archaeological Record of Plant Domestication and Utilization

CHAPTER 1

The Shaping of Modern Paleoethnobotany

PATTY JO WATSON

By "modern" I mean the past thirty years, a period that covers the widespread adoption of flotation/water separation systems for obtaining archaeobotanical remains and the emergence of a generation of paleoethnobotanists who are also anthropologists and archaeologists. This same period is coeval with Dick Yarnell's career, and many members of that younger paleoethnobotanical generation are his students.

It was Dick's work, capitalizing on the vision and opportunity provided by Joe Caldwell and the Illinois State Museum, that helped make our 1960s archaeospeleological exploits in Salts Cave (Watson 1969) legitimate science in the eyes of our archaeological colleagues. Deep cave archaeology had its image problems in the 1960s and 1970s; having as an additional research focus one of the largest repositories of human paleofecal deposits known anywhere did not help matters (Bryant 1974; Marquardt 1974; Watson, ed. 1974; Yarnell 1969, 1974a, 1974b). Dick was a wonderful caving and field companion. Grumbling cheerfully, he made many trips to remote subterranean locales, such as Indian Avenue in Salts Cave, and spent countless hours *après* cave instructing us in ethnobotany and natural history. Some of my most cherished archaeological memories are of trips into Salts Cave and Mammoth Cave with Dick Yarnell, Jean Black Yarnell, and Louise Robbins.

Introduction

Others, including Yarnell (1970), have charted the developmental course of ethnobotany and paleoethnobotany (Bohrer 1986; Ford 1979; Gremillion 1993b; Helbaek 1960, 1963; Miller 1991; Pearsall 1989; van Zeist et al. 1991; see also Chapman and Watson 1993; Wagner 1988; Watson 1976). My intention here is to comment upon the nature of contemporary paleoethnobotany as a distinctive approach to knowledge about the human past in three geographic areas: the Near East, whence came much of the initial impetus for the development of paleoethnobotany as a formal subdiscipline; the Eastern Woodlands of North America, where most of Dick Yarnell's career has unfolded and where some of the most intensive paleoethnobotanical research anywhere in the world is taking place right now; and the southwestern United States, where Dick did his master's thesis research (Yarnell 1965) and where paleoethnobotany—after a couple of decades of semineglect—is finally coming into its own. A common theme in the three regions is the rags-to-riches progression of paleoethnobotany over the past twenty years from peripheral and rather poverty-stricken status to a central place within the archaeological establishment.

A parallel theme is the change in paleoethnobotanical techniques and equipment from homemade plant presses and simple field observation supplemented by the use of discarded microscopes gratefully received from freshman biology labs to gold-plated seeds examined by scanning electron microscopes, $10,000 Swiss optical instruments, and phytolith, isozyme, and DNA analyses, plus routine submission of specimens for accelerator mass spectrometry (AMS) dates at several hundred dollars each.

These are parallel and closely related themes, but I do not think it is the case that paleoethnobotanists are now sought after, respected, and admired simply because they know how to use very expensive high-tech hardware. Quite the contrary, I think what has happened in the past twenty years is that paleoethnobotanists have demonstrated the value of their work so successfully that they have been awarded high status and access to good equipment.

Paleoethnobotany and Near Eastern Prehistory

The pioneer paleoethnobotanist working on Near Eastern materials was Hans Helbaek of the Danish National Museum in Copenhagen. Helbaek was largely self-taught in archaeobotany, yet, so far as I am aware, it was he who coined the word *paleoethnobotany* to mean the identification and cultural interpretation of plant remains from archaeological sites (Helbaek 1960, 1963). He was invited by Robert J. Braidwood (Oriental Institute, University of Chicago) to join the 1954–1955 Iraq-

Jarmo Project expedition as a consequence of his work on charred plant remains and plant impressions obtained during the first full excavation season (1950–1951) at Jarmo in northern Iraq. The 1954–1955 season (when Helbaek was a line item in Braidwood's National Science Foundation grant) marked his formal entry into Near Eastern archaeobotany, a field he dominated for the next twenty years.

Helbaek's scholarship was meticulous and of very high quality, but he was a little defensive about his lack of formal training and of advanced degrees in botany. He sometimes found it difficult to establish easygoing collegial relations with younger investigators who had much more advanced knowledge of genetics, systematics, plant ecology, and agronomy than he did, but who did not have his broad experience and practical mastery of essential lab techniques, many of which he originated.

Helbaek was also a loner who was not especially interested in passing on knowledge and skills to a group of students. He was productive as regards publications, however, and a nonstop worker. Virtually singlehandedly he established the study of macrobotanical remains as an important part of archaeological research in western Asia. He also continued to carry out archaeobotanical studies in Europe and England, but it is his work on plant remains from such key sites and regions as Jarmo (Iraq; Helbaek 1960), Deh Luran (Iran; Helbaek 1969), and Beidha (Jordan; Helbaek 1966) that ensures his position among the immortals of paleoethnobotany.

In the period since Helbaek's death, his one-man research program has been continued and expanded by a vigorous international community of scholars. His successors have, of course, accumulated and interpreted far more detailed evidence than he ever had at his disposal. Although Helbaek employed laboratory flotation prior to its field use by Flannery, Hole, and himself in Deh Luran, Iran, during the mid to late 1960s, he did not live to see the widespread use of flotation by Near Easternists. Powerful impetus for the creation of water separation techniques to obtain botanical materials from various Old World archaeological sites was provided by the school of paleoeconomy Eric Higgs put together at Cambridge University. Out of "the Higgery," as this group was nicknamed, came a widely used flotation/water separation system often called either the "Higgs machine" or the "Cambridge machine" (Jarman et al. 1972).

Developments in paleoethnobotanical hardware in the Near East and the Eastern Woodlands came together in the 1960s and 1970s when the Braidwoods, botanist Bob Stewart, and Stewart's graduate student Bill Robertson incorporated Stuart Struever's flotation approach, developed in west central Illinois, to macrobotanical recovery in their work at the Turkish site of Çayönü (Stewart and Robertson 1973). During the mid to late 1960s, doubtless because of influence from the Higgery, David French was experimenting with water separation systems in his research on prehistoric sites in eastern Turkey (Weaver 1971). Conversations with French at the British Archaeological Institute in Ankara during 1968 and 1970, in addition to discussions with Struever in the United States, spurred my own concern with automated water separation systems. As a result of influence from both French and Struever, Bill Robertson designed and I assembled a Rube Goldberg–type tank flotation system to be used by Bill Marquardt, Louise Robbins, and me in Kentucky in 1974. We called the resulting device the "SMAP (Shell Mound Archaeological Project) machine" (Watson 1976).

Study and interpretation of archaeobotanical remains from the Near East have by now been formalized and institutionalized at a number of places, such as the Biologisch-Archaeologisch Instituut in Groningen (Miller 1991:133) and the Department of Human Environment, Institute of Archaeology, University College, London (see Hillman and Davies 1990). The International Work Group for Paleoethnobotany has been in existence for more than twenty years (van Zeist et al. 1991), and the Sumerian Agriculture Group for ten (Postgate and Powell 1984).

As a result of paleoethnobotanical research during and after the Helbaek era, interested scholars know or can find out what the evidence is for the earliest crop plants in the Near East; where and when they seem to have been domesticated and how they were processed; what wild plants continued to be used; when and where irrigation techniques were developed; what seeds and fruits were favored as funerary offerings at Bronze Age Ur; and much, much more (for recent summaries see Anderson 1992; Miller 1992). In spite of severe problems caused by political instability throughout the region, indomitable paleoethnobotanists continue to analyze and interpret the primary evidence for prehistoric and early historic plant use and agriculture in the Near and Middle East.

Paleoethnobotany and Eastern
North American Prehistory

In contrast to western Asia, ethnobotany in North America was a well-recognized subdiscipline for many decades before paleoethnobotany developed. This relative wealth of ethnobotanical information is obviously an advantage, but it also poses a paradox: given the well-developed interest by anthropologists and others in plant use by indigenous North American groups, why did it take so long for North American archaeologists to pursue ethnobotany into the prehistoric past? A possible conceptual reason is the initial difficulty in establishing meaningful chronologies for North America, where it was not until the mid-1920s that significant time depth was demonstrated. Prior to that period, a common underlying assumption may have been that what was documented ethnographically and ethnohistorically was more or less all there was. Then in the decades immediately subsequent to the landmark Folsom discovery, the majority of archaeological attention was focused quite narrowly on time and space distributions of diagnostic artifacts. Even so, it is rather striking how much discussion there was of agriculture, of horticulture, and especially of maize and how little concomitant systematic effort was applied to obtain empirical information about these matters until the 1970s. Part of the explanation is practical or technical: until some twenty years ago there were no good techniques available to archaeologists for retrieving identifiable charred plant remains from deposits at open sites. Another and more powerful reason, however, is the artifact-centric nature of traditional archaeology. It is only recently that animal bones and plant remains have been admitted as cultural materials on an equal footing with pots, projectile points, and postmolds, a fact emphasized by Stephen Williams at the 1991 Southeastern Archaeological Conference when he presented the Lower Mississippi Valley Survey's coveted C. B. Moore Award to Gayle Fritz for her paleoethnobotanical contributions to southeastern United States archaeology.

Among the founding figures in North American paleoethnobotany, Melvin Gilmore and Volney Jones at the University of Michigan and Hugh Cutler and Leonard Blake at the Missouri Botanical Garden are dominant, not only because of the quality and quantity of their work, but also because of their prominent institutional settings. For several decades, Gilmore and then Jones directed the Ethnobotany Laboratory at the Museum of Anthropology, University of Michigan; Cutler was long-time Curator of Useful Plants at the Missouri Botanical Garden in St. Louis, where Blake also worked for more than twenty years.

Dick Yarnell was a 1963 Michigan doctoral graduate and, like his mentor, Volney Jones, combined ethnobotanical with archaeobotanical research. His dissertation is ethnobotanical (*Aboriginal Relationships Between Culture and Plant Life in the Upper Great Lakes Region*), but most of his more recent work is interpreting ancient uncharred archaeobotanical material as well as flotation-derived charred plant remains from dozens of archaeological sites. In this enterprise he occupies a justifiably eminent position, but he has contributed equally importantly by teaching and encouraging many students, a few of whom have contributed to this volume (namely Gary Crawford, Gayle Fritz, Paul Gardner, Carol Goland, Kristen Gremillion, Julia Hammett, and Gregory Waselkov).

As was true in the Near East, up to about fifteen years ago the number of full-time paleoethnobotanical researchers in and on eastern North America was very small. By the 1980s, however, there were a dozen or more, and the group continues to grow. One factor that helps explain this increase in both places is the highly influential interdisciplinary fieldwork designed by R. J. Braidwood to seek primary evidence for the origins of food production in western Asia. For some twenty years (1950s to 1970s) Braidwood's teams were funded in substantial part by the National Science Foundation, a fact that showcased the research and encouraged emulation, especially by United States–based archaeologists who wanted NSF money. Braidwood was the first Near Easternist since Raphel Pumpelly in the 1904 Anau expedition to mount an explicitly multidisciplinary project that combined natural scientists and archaeologists working together in the field and collaborating on analyses and publication (Braidwood and Howe 1960; Braidwood et al. 1983; Pumpelly 1908).

Another important factor in the blooming of paleoethnobotany in North America is the focus on subsistence, ecology, and economy fostered by the Binfordian New Archeology of the 1960s, which deserves considerable credit for seeing that ecofacts were accorded as much respect as artifacts. By now, many hard-core North American archaeologists willingly, even eagerly, follow the twists and turns of recent and current paleoethnobotanical dramas: for example, (1) the *Chenopodium* drama, when Bruce Smith used scanning electron microscopy to demonstrate which archaeobotanical species and specimens were tame and which were wild (Smith 1984); (2) the flotation-methodological, or poppy-seed, drama, when Gail Wagner's clandestinely administered blind tests showed the strengths and weaknesses of various flotation setups (and of their operators) with results published in *American Antiquity* (Wagner 1982); (3) the Bat Cave drama, when Chip Wills, subsequent to Michael Berry's deconstruction of the ancient maize finds in the Southwest, reexcavated Bat Cave and dated the maize there directly using the AMS radiocarbon technique (Berry 1985; Wills 1988a, 1988b); and (4) the wild-*Cucurbita*-gourds-of-eastern-North-America drama, whose complex plot continues to thicken rapidly about the central question, Did an indigenous wild *Cucurbita* population north of Mexico give rise to many of the modern domestic squashes? A strong case can be made for an affirmative answer to this question, but detailed research is still ongoing (Cowan and Smith 1993; Decker-Walters et al. 1993; Newsom et al. 1993).

In spite of all this excitement as the frontiers of anthropological science are being advanced by paleoethnobotanists, some postprocessualist critiques have targeted what I have identified as a wellspring of paleoethnobotany—Americanist archaeology's "econothink" focus (the adjective is Bob Hall's [1977])—as narrow, distorting, and dehumanizing (e.g., Hodder 1989). The realization that such data may not always be sufficient for explaining cultural history, however, does not detract from the fact that they are always necessary.

By definition, paleoethnobotanists are concerned with the cultural and social contexts—the functions and meanings—of their botanical data. Some of the current practitioners are carrying these concerns into realms of sociopolitical organization, ideology, and gender where most Binfordians declined to tread (Hastorf 1991; Scarry 1988; see also Watson and Kennedy 1991). Thus I believe paleoethnobotany can promote and profit from the best features of both processualist and postprocessualist approaches. Paleoethnobotanists have already put onto a much firmer footing than was previously possible our understanding of when, where, and how agriculture developed in the eastern United States and of the varying significance of the development of agriculture across time and space from Archaic to Woodland to Mississippian and from the Northeast to the Midwest, the Midsouth, the trans-Mississippi South, and the Southeast (Asch and Asch 1985b; Chapman and Shea 1981; Delcourt 1987; Delcourt et al. 1986; Ford, ed. 1985; Fritz 1990;

Smith 1989, 1992b; Wagner 1987; Watson 1985, 1989, 1991a, 1991b;
Yarnell 1986, 1994; Yarnell and Black 1985). We know that people in
the Eastern Woodlands were growing sunflower (*Helianthus annuus*),
sumpweed (*Iva annua*), chenopod (*Chenopodium berlandieri*), maygrass
(*Phalaris caroliniana*), gourd/squash (*Cucurbita pepo*), and bottle gourd
(*Lagenaria siceraria*) 2,500 years ago. We have strong grounds for thinking
that every one of these plants was domesticated indigenously north of
Mexico and independent of early agriculture there. We know, too, that
maize appears late in the East (Chapman and Crites 1987; Riley et al.
1994) and is not a causal factor in the development of Mississippian (or
other) social complexity.

As paleoethnobotanical inquiry in the Eastern Woodlands continues, we will also come to understand why agriculture arose or was intensified in some places and not in others. Adequate explication of that
why will require, as it always does, thoughtful consideration of a wide
array of potential explanatory factors, including sociopolitical and ideo-
logical processes (possibly relevant in some parts of the Eastern Woodlands [e.g., Bender 1985; Brown 1985]) as well as the environmental
imperatives that seem most cogent for the Levantine Near East (Bar-
Yosef and Belfer-Cohen 1989, 1992; Henry 1989).

Paleoethnobotany and Prehistory
in the Southwestern United States

The Southwest provides yet another major theme in the paleoethnobotanical saga. Like the eastern United States, the Southwest is
blessed with numerous dry rockshelters and small caves where uncharred plant remains are beautifully preserved in stratigraphic context. Seeming triumphs some forty years ago in revealing the history
of maize from such settings have recently been thoroughly reevalu-
ated and reinterpreted often with the aid of AMS dating (Berry 1985;
Matson 1991; Wills 1988a, 1988b). New evidence and new understand-
ings of old evidence about ancient plant use are emerging, but with
few exceptions (e.g., Fish 1989; Minnis 1985a, 1985b, 1992) archaeologists theorizing about early agriculture in the Southwest focus intensively, almost obsessively, upon maize. Certainly maize was extremely
important, but it is equally certainly not the whole story of plant use
by indigenous populations in the Southwest, nor is it even the whole

story of indigenous agriculture there. What that whole story might be is not yet known in detail, but evidence from paleoethnobotanists participating in recent Cultural Resource Management projects is beginning to provide some hints. For example, it appears that the Hohokam cultivated not only maize, but also little barley (*Hordeum pusillum*), chenopod (*Chenopodium* sp.), amaranth (*Amaranthus* sp.), panic grass (*Panicum sonorum*), cholla (*Opuntia* sp.), and agave (*Agave* sp.) (Fish 1989:46; for recent summaries of southwestern archaeobotany see Ford 1985; Fritz 1994; Matson 1991; Minnis 1985a, 1985b, 1992; Wills 1988a, 1988b). Maize was, however, present in the Southwest by ca. 3000 B.P. and was central to some economies by 2500 B.P. in a way that was not the case in the East until a millennium and a half later. On the other hand, domesticated food plants (sunflower, sumpweed; perhaps also chenopod) were present a few to several hundred years earlier in the East than in the Southwest, and the container crops *Cucurbita* and *Lagenaria* were probably domesticated independently in the East, perhaps as early as in Mexico (Cowan and Smith 1993; Decker-Walters et al. 1993; Newsom 1988; Newsom et al. 1993; Smith 1992b).

Discussion and Conclusions

By way of capsule summary, one could say that legitimization—the move from periphery to center—of paleoethnobotany has proceeded at different rates and along different trajectories in the three areas I have touched upon. Braidwood's determination to find the primary evidence for agricultural origins in the Near East initiated a very productive era of paleoethnobotanical research there in the 1950s that continues to the present. In the Eastern Woodlands of North America the paleoecological, subsistence-settlement system orientation of New Archeology in the 1960s–1970s fostered similar concern with biological remains, and, of course, as regards agricultural origins, especially with plants. In the Southwest, intermittent attention to macrobotanical material in dry caves and rockshelters—often maize and often erroneously dated—has finally given way to systematic wide-spectrum archaeobotanical retrieval, analysis, and interpretation, as in the other areas. The Flotation Revolution of the 1960s–1970s, pioneered by Stuart Struever in the United States and by the Higgery in the Old World, had a dramatically positive impact on all three regions. Retrieval systems, analytical modes,

Patty Jo Watson

and systematization in general have been rapidly evolving in all three places for the past twenty years. Even more important than archaeobotanical machinery and quantification techniques, however, are the dedicated scholars, like Dick Yarnell, whose work has brought paleoethnobotany to its justifiably prominent position and who have educated a new generation of eager and sophisticated researchers.

CHAPTER 2

New Perspectives
on the Paleoethnobotany
of the Newt Kash Shelter

KRISTEN J. GREMILLION

On December 17, 1935, William S. Webb of the University of Kentucky shipped a package to the Ethnobotanical Laboratory at the University of Michigan. Volney Jones, then an assistant to the Laboratory's director, Melvin Gilmore, eagerly took on the task of analyzing this dry vegetal material from an archaeological site in Menifee County, Kentucky, known as Newt Kash Hollow shelter. The size, quality, and significance of the collection rivaled those of similar material from the Ozarks then being studied by Gilmore, prompting Jones to write that "from an ethnobotanical point of view, the cave and shelter area from Kentucky to Arkansas is by far the most interesting and promising in eastern North America" (letter to W. S. Webb dated August 22, 1936; Ethnobotanical Laboratory files, Museum of Anthropology, University of Michigan [UMMA-EL]). After receiving Jones's report, Webb was both pleased at the quantity of information obtained and regretful at having previously neglected this source of data: "I have stirred through Indian beds and shoveled out bushels of 'trash' which in my ignorance I regarded as valueless," he wrote, adding, "I now know that I have probably destroyed a large body of valuable information" (letter to V. Jones dated August 28, 1936; on file, UMMA-EL).

The history of research on botanical materials from the Newt Kash Shelter is the story, in microcosm, of the evolution of paleoethnobotanical method and theory over the past sixty years. Extraordinarily well-preserved plant materials from this site provided inspiration for the hypothesis of premaize agriculture in the Eastern Woodlands of the United States and produced key empirical support for its acceptance.

23

Since the time of Volney Jones's initial observations on the distinctive morphology of certain seeds and their stratigraphic position within the shelter, Dick Yarnell and his colleagues have worked to refine our understanding of both the chronological placement of the Newt Kash remains and the evolutionary significance of the patterns of morphological variation they display. Data obtained from Newt Kash and nearby rockshelters clearly illustrate the value of improvements in field and laboratory processing and analysis that have occurred in paleoethnobotany since Jones's original study, as well as the need for further refinement of methods and techniques.

Seed Morphology, Site Chronology, and the Origins of Agriculture

Along with data from the Ozarks (Gilmore 1931), the Newt Kash materials were to provide the initial empirical basis for the developing notion that eastern North America was an independent center of plant domestication. Two lines of evidence, seed morphology and site chronology, were to provide much of the support for this hypothesis over the coming decades (Fritz 1990; Gremillion 1993b; Smith 1989). The ways in which these sources of information were explored by researchers engaged in the study of materials from Newt Kash document the maturation of paleoethnobotany as a discipline.

The Study of Crop Morphology

Initial nonquantitative observations of seed morphology were eventually replaced by systematic measurement, resulting in the eventual establishment of morphometric data bases for several taxa. However, such studies remained restricted to features of gross morphology visible to the naked eye or under low magnification. Because it was easily recognizable using these techniques, and was known to vary under domestication, seed size was the character initially seized upon in the search for morphological correlates of plant husbandry. New perspectives on the cultural significance of crop morphology were, however, acquired when scanning electron microscopy was applied to the study of archaeological seeds. Morphological data obtained using these techniques proved to be particularly powerful when combined with ecological principles to predict the effects of natural selection upon plants in human-modified habitats.

The Paleoethnobotany of Newt Kash Shelter

As Jones observed in the report he prepared for Webb (Jones 1936), seeds of sumpweed (*Iva annua*) and sunflower (*Helianthus annuus*) from Newt Kash were significantly larger than those of their wild relatives. Although Jones noted this characteristic and described its significance, he did not report any dimensions. Morphological variation in chenopod (*Chenopodium*) was also recognized without the aid of formal measurement. Jones noted both small seeds of *Chenopodium* and larger specimens that he believed might represent a domesticate.

Later researchers greatly increased the interpretive value of Jones's work by measuring the lengths and widths of archaeological seeds and fruits. For example, Yarnell (1972, 1978) was able to demonstrate a significant (if not always smooth) increase in size beginning as early as the Late Archaic by compiling measurements of sumpweed and sunflower achenes and seeds from Newt Kash and a number of other sites. The presence of entire sunflower heads at Newt Kash allowed Charles Heiser to propose additional details of the domestication process, such as development of forms intermediate between the giant monocephalic cultigen and the wild branched sunflower (Heiser 1978). For some time, it seemed that only a weak case could be made for domestication of *Chenopodium* (Asch and Asch 1977). Although they recognized several differences in morphology between dry shelter chenopods and wild populations, David and Nancy Asch concluded that traits such as thin seed coats fell within the normal range of variation of wild populations. Size difference appeared to have been eliminated as a criterion for domesticated status in *Chenopodium* when the larger seeds from Newt Kash were shown to actually be those of poke (*Phytolacca americana*), whose seeds are morphologically similar but larger (Asch and Asch 1977). Rectification of this error was only possible through close scrutiny of seeds and comparison with published information and botanical specimens. The benefits of periodic reanalysis of plant remains using improved observational techniques are also illustrated by Richard Ford's revision of identifications of *Lagenaria siceraria* (bottle gourd) and *Cucurbita* (gourd/squash) rind and seeds from the Newt Kash collection (Ford 1986).

Further revision of opinions regarding the domesticate status of *Chenopodium* by subsequent researchers clearly illustrates how systematic formulation of the morphological consequences of selection under domestication combined with sophisticated observation techniques aided the study of agricultural origins. These consequences included reduction of the seed coat and a change in seed cross-sectional shape from rounded to rectanguloid with truncate margins (Smith 1984, 1985a, 1985b; Wilson 1981). Observation and measurement of multiple seed characters have since allowed for even more detailed description and analysis of archaeological collections of *Chenopodium*. At the same time, the growing morphometric data base for this taxon has increased awareness of the range of morphological variation likely to occur within archaeological samples of crop seeds. One aspect of this variation has been explored through examination of archaeological and ecological relationships between domesticated and wild-type *Chenopodium* (Gremillion 1993a).

The potential of these techniques for the study of crop morphology is demonstrated by recent studies of a small collection of *Chenopodium* seeds from Newt Kash. I examined two lots of seeds, one of which, extracted from a fecal sample, bears the same catalog number (UMMA-EL 1114) as specimens that produced an accelerator radiocarbon date of 3400 ± 150 B.P. (Smith and Cowan 1987). The other had provided seeds for the earlier (1977) study by David and Nancy Asch (UMMA-EL 1216). No provenience information is available for either sample.

Interestingly, seeds from the two samples were quite different in appearance. The seeds from paleofeces, associated with the early date, were all dark to translucent reddish in color and were exceptionally large (nearly 2 mm in diameter) (figure 2.1). In addition, these specimens had seed coats less than 20 microns thick (Smith's "baseline" average for domesticate status) (Smith 1987b). None had clearly truncate margins or completely smooth seed coat surfaces (both characteristics initially identified by Smith [1985a] as hallmarks of domesticate status). The other sample, UMMA-EL 1216, was represented by smaller seeds, also dark in color but with clearly truncate margins. Among them were two atypical specimens that resembled wild chenopod seeds, with relatively thick seed coats and rounded margins.

Some of the differences observed may be artifacts of distortion in the paleofecal seeds. However, the strong possibility also exists that the two collections represent temporally and/or spatially separated populations and thus may reflect distinct patterns of crop evolution and human/plant interaction. Unfortunately, without proper contextual information these possibilities cannot be fully explored. However, this comparison of *Chenopodium* seeds from Newt Kash points out that de-

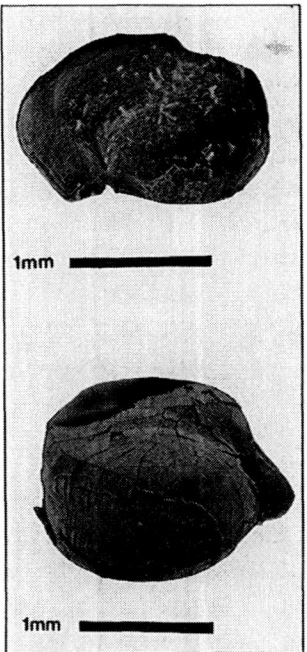

Figure 2.1. *Chenopodium* seeds from Newt Kash. *Top,* thin-testa (domesticate type) seed from UMMA-EL 1114 viewed at 44.8 × magnification; note textured seed coat and rounded seed margins. *Bottom,* thin-testa seed from UMMA-EL 1216 viewed at 43.1 × magnification; note truncate seed margin and relatively smooth seed coat.

tailed studies of seed morphology are often rewarded by insights that a cursory examination would not lead us to suspect are possible.

Culture Change and Chronology: Then and Now

The other critical line of evidence for agricultural origins in eastern North America, site chronology, followed a similar pathway in the direction of increasing precision and refinement of techniques. Application of the principle of superposition was eventually supplemented by radiocarbon dating of associated deposits and, ultimately, of the cultigen remains themselves. Late Archaic *Cucurbita* rind from sites in the Midcontinent was hailed as evidence that agriculture was ultimately of Mesoamerican inspiration (Chomko and Crawford 1978), a conclusion that carried embedded within it the assumption that the earliest crop plant provided the template for all subsequent agricultural traditions. The direct dating of Middle Holocene *Cucurbita* remains from Illinois using the accelerator mass spectrometry (AMS) technique (Asch and Asch 1985b; Conard et al. 1984) seemed to offer additional support for a Mesoamerican origin for plant husbandry in the Eastern Woodlands. However, this position was soon severely weakened by reassessments of the taxonomy and morphology of *Cucurbita pepo*. Middle Holocene *Cucurbita* rinds and seeds from the East were found to be morphologically indistinguishable from those of the Texas wild gourd (*C. pepo* ssp. *ovifera* var. *texana*). If, in fact, both represent a lineage of indigenous eastern North American wild *Cucurbita* gourds (Cowan and Smith 1993; Decker-Walters et al. 1993; Smith et al. 1992), the Middle Holocene material need not imply either crop diffusion from Mesoamerica or cultivation of domesticated forms.

Although accurate chronology remains indispensable to the interpretation of past subsistence change, temporal placement is in the 1990s both easier to come by and less satisfying as an end product than it was in the 1930s. It was implicit in the archaeological literature of Jones's time that the important questions about agricultural origins would be automatically answered along with the determination of whether Mesoamerican or native crops were utilized earlier (Gremillion 1993b). The cucurbits (*Lagenaria* and *Cucurbita*) played a key role in this debate because they appeared to predate maize (*Zea mays* ssp. *mays*) in many archaeological contexts and were assumed to be both domesticated and of Mesoamerican origin. Thus the temporal priority of cucurbits, if it could be established, would ensure a Mesoamerican "inspiration" for agriculture and thereby strike a deadly blow to the independent domestication hypothesis.

Evaluation of the relative significance of the native and imported agricultural traditions represented at Newt Kash initially relied upon application of the principle of superposition. The stratigraphic position of the cucurbit remains could not be determined with any certainty, but the remains did appear in levels below those containing maize. Jones acknowledged that the "startling conclusion that agriculture had a separate origin in the bluff shelter area" hinged upon the then-undetermined chronological relationships between cucurbits and the native seed crops (Jones 1936:163). Thus although both cucurbits and maize were viewed as evidence of contact with Mesoamerica, an argument could be made for an independent indigenous agricultural tradition that had arisen at least somewhat independently of influences from the south.

Additional support for this argument was provided by the fact that cucurbit remains, like those of maize, were relatively scarce in the sample examined by Jones (although he later complained to Webb that his discussion of quantities was flawed because he mistakenly believed the sample he had received represented the total recovered from the site) (letter from Jones to Webb dated July 16, 1937; on file, UMMA-EL). Thus he concluded that tropical cultigens were of little importance to site inhabitants.

The radiocarbon dating technique became available to archaeologists in the 1950s. Curated materials from Newt Kash were among the earliest analyzed by the University of Michigan Radiocarbon Laboratory. These dates (2650 ± 300 B.P. and 2600 ± 300 B.P.) (Crane 1956) were obtained on remains of grasses from the site using the solid carbon method, which has since been abandoned in favor of more accurate gas and liquid scintillation techniques (Gowlett 1987). The advent of radiocarbon dating using the AMS technique solved many problems of temporal control by allowing for the direct age assessment of cultigen seeds and fruits. For example, the direct accelerator date of ca. 1500 B.C. (3400 ± 150 B.P.) later obtained on cultigen *Chenopodium* from Newt Kash pushed back estimates for the initiation of plant husbandry well into the second millennium B.C. (Smith and Cowan 1987). More recently, consumption of domesticated sunflower by people who used Newt Kash around 1000 B.C. has been confirmed by an accelerator date on paleofeces from the site (3025 ± 55 B.P.: 1075 B.C.; Beta 62664, ETH 10491) (Gremillion 1994). Cucurbits from Newt Kash have not been directly dated.

Paleoethnobotanical Data Recovery and Analysis

The sixty-year history of research on plant remains from Newt Kash is also a case study in the development of techniques for acquiring and analyzing the bulk samples of plant remains that have become the mainstay of modern paleoethnobotanical research. Excavation techniques employed in eastern Kentucky rockshelters during the early part of this century did not always preserve the distinctions between different deposits that are so critical to interpreting the significance of artifactual and subsistence remains. Botanical remains were, sadly, often regarded as refuse by these excavators. In contrast, fine-screen recovery techniques are now standard in the field and are applied as a matter of course on rockshelter sites similar to Newt Kash. Laboratory methods developed since Jones's early study also allow more refined understanding of the subsistence importance and economic role of various plant foods.

Excavation

Although morphology and chronology have played the most important roles in documenting premaize agriculture at Newt Kash and other sites, our knowledge of the development of agriculture would be greatly impoverished without simultaneous improvements in methods of recovering plant remains and quantifying the resulting data. The published record of the excavation of Newt Kash speaks eloquently on this point: "Excavation was begun," report Webb and Funkhouser, "at the east end of the shelter by the usual method of starting at the outside edge and shoveling the ashes and sand down the incline and working backwards toward the rear wall in a twenty foot cut" (Webb and Funkhouser 1936:112). Fortunately, pits were excavated carefully and their dimensions recorded. However, if any of the plant materials were recovered from these contexts, it has not been recorded; provenience information in the UMMA-EL files is limited to "upper layer," "lower layer," and "bedding."

Although we cannot apply improved excavation and recovery techniques to the fieldwork of half a century ago, we can use them consistently, and with an eye to improving and refining them, in the future. Fortunately, many dry shelters remain, though few are undisturbed by illegal or irresponsible digging. However, the work of Wes Cowan and Cecil Ison, with that of others (Cowan 1978a, 1978b, 1979a, 1979b, 1985a, 1985b; Cowan et al. 1981; Gremillion 1993c; Ison 1988; O'Steen et al. 1991), has clearly demonstrated that the potential of these sites for documenting prehistoric subsistence change is increased manyfold by careful excavation of complex sediments coupled with fine-screen recovery of plant remains.

Analysis, Quantification, and the
Reconstruction of Subsistence Patterns

Unlike new excavation techniques, improved methods for examining and quantifying mixed samples of plant remains can be applied to material from previously excavated sites such as Newt Kash. Unfortunately, rockshelter collections are seldom subjected to the kinds of standard quantification procedures pioneered by Dick Yarnell (1974a, 1974b) and refined over the years. This abandonment of standard protocol when rockshelter collections are the subject of study may be in part a response to deprivation. Like people grown accustomed to food scarcity, data-starved paleoethnobotanists may be tempted to abandon judicious management practices when faced with a glut of resources. As a result, the overall composition of collections of well-preserved plant remains is often neglected in favor of more specialized analyses of cultigen morphology. Perhaps the sheer volume and diversity of rockshelter collections are unusually daunting for researchers. Ironically, carbonized samples, which are more severely impacted by processes of decay, are also more likely to be subjected to careful quantification procedures.

Another reason analysts are reluctant to place much faith in quantification of botanical residues from rockshelters is that they were not systematically sampled in the field. In fact, contextual information of any kind generally ranges from minimal to absent. Thus in these cases detailed interpretation is hampered by lack of information about many key variables (such as sample size and source) that are routinely controlled in modern excavation and analysis. However, neglecting to apply the analytic techniques at our disposal to collections that share these deficiencies, although it eliminates certain methodological difficulties, takes interpretive caution to unnecessary extremes. Instead, I argue that the strategy of applying sound techniques to poorly documented collections can actually increase the research potential of well-preserved but poorly collected samples of plant remains.

An analysis of the general botanical composition of a sample of material from Newt Kash was conducted in order to demonstrate the value of such an approach. The purpose of this analysis was twofold. First, no bulk samples of botanical material from the site had ever been quantified by past researchers. Because quantified data in the form of counts, weights, and ratios form the basis for interpretations of subsistence by paleoethnobotanists, absence of such data in the case of this important site seemed a serious omission, despite the contextual and sampling problems that Newt Kash shares with most collections of plant remains from dry shelters (see also Fritz, this volume). By comparing these data with those obtained from a similar dry shelter in eastern Kentucky, the Cold Oak shelter, I hope to place the Newt Kash paleoethnobotanical record in regional context as one example of a widespread pattern of subsistence change. A second goal of this analysis was to use the results obtained from these two sites to illustrate the benefits of the many improvements in technique and method that have become available since 1936. Thus the analysis of plant remains from Newt Kash is also intended as a reminder of how much more closely the scientific potential of this important site might have been realized had it been excavated in the 1990s instead of the 1930s. This is done not to diminish the efforts of past researchers, but as a reminder to those of the present and future to seek further improvements.

With the cooperation of Richard I. Ford, Director of the Paleoethnobotany Laboratory at the University of Michigan, I was able to arrange for the loan of an unsorted and uncataloged sample of plant material from the site. The original UMMA tags, apparently in Volney Jones's handwriting, were still in the boxes containing the material. Although Jones may have removed some of the larger objects (C. W. Cowan, personal communication 1992), a wide range of sizes and types of botanical material remained. This sample, approximately 3 l in volume, was sifted through a series of geological sieves. Because of its considerable bulk, only material greater than 2.6 mm in size was completely sorted into taxonomic categories. However, all seeds and unusual items greater than 0.5 mm in diameter were identified and counted.

A great variety of materials was encountered. Lithic debitage and sandstone fragments made up the bulk of the nonbotanical materials. Small quantities of animal bone, feathers, and mussell shell were also identified. Plant food remains included nutshell (primarily thick-shelled hickory [*Carya* spp.], but also chestnut [*Castanea dentata*], acorn [*Quercus* spp.], walnut [*Juglans nigra*], and butternut [*Juglans cinerea*]) and honey locust (*Gleditsia triacanthos*) pods in addition to seeds and fruits of cultigens, weeds, fleshy fruits, and trees. Fibers, grass stem fragments, short lengths of giant cane (*Arundinaria gigantea*), and moss leaves were also observed (table 2.1).

Screening and Seed Recovery.

Jones (1936) identified several taxa represented by seeds and fruits, wood charcoal, and fibers. However, the number of seed types appears to be higher in the recently analyzed sample. I was able to identify some twenty-five seed taxa, including identifiable but unknown taxa,

The Paleoethnobotany of Newt Kash Shelter

Table 2.1. Newt Kash Quantified Sample, Plant Remains

Category	Weight (g)	%
Grass stems	16.00	5
Leaves and pedicels	2.98	1
Fibers and epidermis	60.05	18
Bark	0.70	tr[1]
Cane (*Arundinaria gigantea*)	2.21	1
Uncharred woody stems	51.22	15
Wood charcoal	31.72	9
Unknown	26.93	8
Moss leaves	0.22	tr
Plant food remains	151.64	44
Thick-shelled hickory (*Carya* sp.)	100.59	29
Thin-shelled hickory	3.38	1
Hickory husk	5.42	2
Acorn shell (*Quercus* spp.)	14.51	4
Chestnut shell (*Castanea dentata*)	6.43	2
Walnut shell (*Juglans nigra*)	18.12	5
Butternut shell (*Juglans cinerea*)	1.58	1
Maygrass glumes (*Phalaris caroliniana*)	0.11	tr
Honey locust pod (*Gleditsia triacanthos*)	0.71	tr
Cucurbita rind	0.03	tr
Lagenaria rind	0.01	tr
Seeds	0.75	tr
Total	343.67	101[2]

[1]tr = trace (less than 0.5%).
[2]Does not total to 100% because of rounding error.

as well as specimens identified at least to the taxonomic level of family (table 2.2). These included very small (ca. 0.8 mm) reddish seeds that were irregularly circular in transverse section (designated "minute unknown"). The varied assemblage of grass grains included little bluestem, crabgrass-like specimens, foxtail grass, an abundant unidentified large-seeded species (10 to 14 mm in length), and "Type E," of which the palea and lemma were 7 to 9 mm long, translucent, and prominently veined. These were likewise not reported by Jones, who mentioned only about thirteen seed types in the 1936 publication. Thus this reanalysis

Table 2.2. Newt Kash Quantified Sample, Seed Counts and Percentages

Category/Taxon	Count	%
Crops and possible crops		
Maygrass[1] (*Phalaris caroliniana*)	30	10
Chenopod[1] (*Chenopodium berlandieri*)	18	6
Sumpweed[1] (*Iva annua*)	31	11
Sunflower[1] (*Helianthus annuus*)	8	3
Giant ragweed[1] (*Ambrosia trifida*)	1	tr[2]
Knotweed (*Polygonum* sp.)	1	tr
Amaranth (*Amaranthus* sp.)	1	tr
Total	90	30
Grasses (Poaceae)		
Large grooved	54	19
Crabgrass-like (*Digitaria* sp.)	15	5
Foxtail grass (*Setaria* sp.)	1	tr
Little bluestem (*Andropogon scoparius*)	9	3
"Type E"	31	11
Misc. unknown	10	4
Total	120	42
Fleshy fruits		
Sumac (*Rhus* sp.)	1	tr
Bramble (*Rubus* sp.)	1	tr
Total	2	tr
Trees		
Tulip tree (*liriodendron tulipifera*)	5	2
Other		
Beggar ticks (*Bidens* sp.)	2	1
Beggar lice[1] (*Desmodium* sp.)	1	tr
Bulrush (*Scirpus* sp.)	1	tr
Buttercup (*Ranunculus* sp.)	1	tr
"Minute unknown"	70	24
Aster family (Asteraceae)	2	1
Total	77	26
Total seeds	294	100

Note: Unidentifiable seeds and seed fragments are not reported.
[1]Denotes taxa identified by Jones (1936).
[2]tr = trace (less than 0.5%).

illustrates that systematic screening greatly increases the likelihood of obtaining a representative sample of seeds, although it is difficult to assess the difference between previous and present studies in this respect because it is uncertain whether all identifications were reported or only those deemed particularly significant. Analytic methods were also not described in the 1936 report.

Quantification and the Role of Food Production.

Like systematic processing, calculation of simple ratios and percentages has become a standard technique for estimating the subsistence importance of plant foods. Of course, diet composition cannot be precisely assessed by directly comparing quantities of different classes of plant remains, which vary in preservation potential and the amount of food that each represents (Miller 1988; Popper 1988). However, differences in the relative importance of plant foods in different prehistoric populations can be evaluated by comparing carefully selected statistics (such as seed and nutshell percentages) that describe the patterning of occurrence of plant food remains. Such comparisons have greatly assisted the reconstruction of long-term subsistence trends (including the evolution of food-producing economies) in areas such as the American Bottom (Johannessen 1984), west-central Illinois (Asch and Asch 1985b), and eastern and central Tennessee (Chapman and Shea 1981; Crites 1991).

Unfortunately, differential preservation is a serious impediment to comparison of these large data sets with information from dry shelter sites such as Newt Kash. Plant remains of all types will be more abundant in a dry rockshelter than in an open site occupied by the same group of people engaged in the same activities. Thus comparisons between ethnobotanical assemblages from dry rockshelters, in which preservation conditions may be considered roughly equivalent, are more reliable than those carried out between sites with different preservation potential.

In the case of Newt Kash, the best candidate for such a comparison is the Cold Oak shelter, located about 25 km southeast of Newt Kash at the head of a small hollow northeast of the Kentucky River (figure 2.2). Organic materials (including cordage, basketry, and wooden implements, as well as food remains) have been exceptionally well preserved there. Fortunately, the employment of careful excavation (O'Steen et al. 1991) has ensured that the value of the paleoethnobotanical as-

Figure 2.2. Map showing locations of key rockshelter sites in eastern Kentucky.

semblage from this site greatly exceeds that of Newt Kash. The primary occupation of this site occurred between ca. 3000 and 2300 B.P., with both Terminal Archaic and Early Woodland components represented in subsurface deposits. Terminal Archaic activity at Cold Oak was associated with an extensive layer containing uncarbonized organic debris (Zones III and IIIa) as well as several features. Botanical remains from Zones III and IIIa produced uncalibrated radiocarbon dates of 2930 ± 70 B.P.:980 B.C. and 2830 ± 60 B.P.:880 B.C., respectively (Gremillion 1993c; Ison 1988). A third Terminal Archaic date (2900 ± 100 B.P.:950 B.C.) was obtained directly on a thick (approximately 3.5 mm) *Cucurbita* rind fragment from Zone III using the AMS method (Gremillion 1993c). Early Woodland deposits were associated with several subterranean storage features, including one grass-lined pit (Feature 2) containing abundant seeds of domesticated chenopod. Charcoal from Feature 2 produced a beta decay date of 2210 ± 60 B.P.:260 B.C. and chenopod seeds were AMS dated at 2590 ± 90 B.P.:640 B.C. (Gremillion 1993c; Ison 1988; O'Steen et al. 1991). Crop remains (including chenopod, maygrass [*Phalaris caroliniana*], sunflower, sumpweed, and cucurbits) were markedly more abundant in Early Woodland contexts, although they were present in Terminal Archaic deposits as well (Gremillion 1993c; O'Steen et al. 1991).

Because Cold Oak closely resembles Newt Kash in environmental setting and primary period of use, data derived from it provide a useful supplement to the poorly provenienced collection from Newt Kash. The percentage of crop seeds (here including weedy grain-bearing annuals that may have been cultivated as well as morphologically altered domesticates) in the Terminal Archaic component at Cold Oak (29%) is considerably lower than the 86% calculated for Early Woodland contexts at the same site (Gremillion 1993c). Crop seeds as defined here make up 31% of identifiable seeds in the quantified sample from Newt Kash (figure 2.3) (if the "minute unknown" seed category is excluded from seed totals, the percentage of crop seeds at Newt Kash rises to 69%).

While it would not be appropriate to claim a temporal correspondence based on sample composition (particularly given the uncertain history of the Newt Kash material), this comparison shows that detailed consideration of quantified data might have allowed the discrimination of contrasting patterns of plant use within the shelter, had adequate information been available. Although Newt Kash is widely known for its spectacular finds of whole sunflower heads and bundles of maygrass, these unusual items appear to be representative of only one portion of the site's occupational history, which reflects the incipient plant husbandry of the Late Archaic as well as the more economically significant farming of the Woodland period.

Evidence from the Cold Oak shelter thus both reiterates and clarifies certain patterns of plant use and management that are evident but poorly understood and defined on the basis of the data from Newt Kash. First, small-scale gardening was practiced in eastern Kentucky during the fourth millennium B.P. and resulted in morphological changes in certain populations of several plant species. This inference is supported

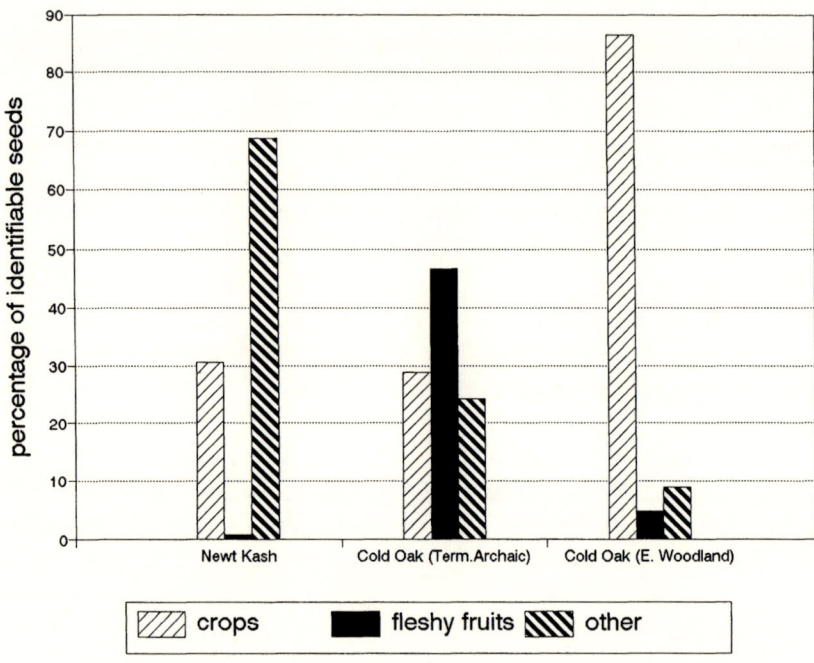

Figure 2.3. Comparison of seed assemblages from Newt Kash and Cold Oak by economic category. The category "crops" includes weedy annuals that may have been cultivated as well as morphologically distinctive domesticates and other taxa with well-documented crop status. For assignment of taxa to economic categories, see table 2.2. Additional taxa identified at Cold Oak but not at Newt Kash are classified as follows: "fleshy fruits" includes elderberry (*Sambucus* sp.), grape (*Vitis* sp.), huckleberry (*Gaylussacia* sp.), black gum (*Nyssa sylvatica*); "other" includes panic grass (*Panicum* sp.), cut-grass (*Leersia virginica*), dogwood (*Cornus florida*), hop hornbeam (*Ostrya virginiana*), maple (*Acer* sp.), naiad (*Naja guadalupensis*).

both by the direct-dated domesticated chenopod from Newt Kash and by examples of domesticated *Cucurbita,* sumpweed, and chenopod from dated contexts at Cold Oak. Late Archaic plant husbandry may also be reflected in bulk samples of plant remains, although these cannot be assumed to represent a single occupation or even component. Second, although Early Woodland occupants of Cold Oak grew a similar set of crops, they stored their harvests in subterranean pits, unlike their Terminal Archaic predecessors. This practice, and the increasing emphasis on garden products that seems to have accompanied it, resulted in the deposition of large numbers of crop seeds in archaeological contexts. The Newt Kash evidence may reflect a similar trend, with the quantified sample and the direct-dated chenopod possibly representing an early phase of plant husbandry. Numerous storage pits (Webb and Funkhouser 1936) and perhaps the bulk of cultigen remains are associated with the later integration of food production with the storage of plant products. Investigations at other nearby dry rockshelters, such as Cloudsplitter, generally support the scenario of initial domestication of weedy annuals during the Late Archaic followed by the concomitant development of food production and crop storage (Cowan 1985a, 1985b, this volume; Cowan et al. 1981).

Cold Oak and Newt Kash: A Study in Contrasts.

Comparison of the paleoethnobotanical data base from Cold Oak and Newt Kash clearly illustrates the scientific potential of such dry rockshelter sites and the different extent to which that potential has been realized over the past sixty years. Morphometric analyses of crop remains from Cold Oak have focused on *Chenopodium,* which is the most abundant seed type in Early Woodland contexts. Documentation of temporal trends in morphological variation would have been impossible without the temporal control offered by the series of radiocarbon dates from the site. By comparison, the value of the Newt Kash chenopod for such studies is quite limited. Although more dates are needed, Cold Oak has contributed toward the documentation of the initial periods of plant husbandry in eastern North America by producing Terminal Archaic examples of sumpweed, maygrass, *Cucurbita,* and *Chenopodium.* Carefully controlled excavation at Cold Oak has ensured that the temporal placement of key plant remains that have not been directly dated can be inferred with some degree of confidence on the bases of provenience and indications of postdepositional disturbance. Features related to different components have been isolated and their functions evaluated. In contrast, we have no hope of ever knowing where the Newt Kash botanical remains were located within the site either horizontally or vertically. Finally, plant remains from Cold Oak were obtained using fine screening of soil samples. Analysis involved careful quantification of plant remains in ways that enable intersite and intrasite comparisons. This procedure has allowed the comparison of plant remains assemblages related to the two primary components on the site, making possible further characterization of two stages in the development of plant husbandry.

Summary and Conclusions

The present analysis has both enhanced and benefited from the application of current analytic techniques to more recently excavated material. New quantitative information from Newt Kash contributes to the growing body of evidence for the initiation of small-scale food production in eastern Kentucky during the fourth millennium B.P. At the same time, new data from sites such as Cold Oak have transformed the significance of Newt Kash; the paleoethnobotanical assemblage of this site can now be examined as an exemplar of a regionwide tradition rather than a singular and cryptic record of early farming.

Unfortunately, their uniqueness of these exceptionally well preserved collections as archaeological phenomena has fostered the belief that they represent similarly atypical kinds of subsistence behavior. In fact, there is no reason we should choose carbonized assemblages as offering a less biased reflection of past food procurement systems; instead, it seems more reasonable to begin with the assumption that the best-preserved, most complete assemblages provide a superior evidential base. Of course, we have yet to devise adequate methods for determining subsistence importance on the basis of relative quantities of food remains. Thus the precise role of agriculture for the occupants of Newt Kash and similar sites remains somewhat obscure. However, the solution to these and other problems of paleoethnobotanical interpretation is not facilitated by disregarding our most valuable sources of evidence on the grounds that they are aberrations.

These results emphasize that even though paleoethnobotany has come a long way empirically, theoretically, and methodologically since 1936, there is still room for much progress in these areas. The rockshelters of eastern Kentucky are likely to continue to play a prominent role both in documentation of agricultural origins and in development of paleoethnobotanical techniques and methods (see for example Cowan, this volume). Although a great deal of professional excavation has been done in the area since 1936, continued investigations are urgently needed to mitigate the effects of ongoing destruction of these important resources. As Raymond Thompson, then Curator of the Museum of Anthropology at the University of Kentucky, replied to Volney Jones's request for additional information on maize from Kentucky shelters, " . . . the only way to get any control material on the eastern Kentucky rock shelters is to get out and dig something new" (letter from Thompson to Jones dated January 26, 1956; on file, UMMA-EL). Thanks to the efforts of scholars such as Dick Yarnell, this task can be accomplished in ways that maximize the scientific potential of these unique and valuable paleoethnobotanical resources.

CHAPTER 3

A Three-Thousand-Year-Old Cache of Crop Seeds from Marble Bluff, Arkansas

GAYLE J. FRITZ

Approximately 3,000 years ago, at least one group of people living in the southern Ozarks chose a dry crevice in a rockshelter beside a small stream, today called Mill Creek, as a storage place for seed stock. The bags of seeds they buried in the crevice demonstrate that, in addition to whatever wild plants and animals these people ate, a portion of their sustenance was derived from domesticated plants. The crops include a type of gourd or squash (*Cucurbita pepo* ssp. *ovifera*), a thin-testa cheno-pod (*Chenopodium berlandieri* ssp. *jonesianum*), sumpweed or marsh elder (*Iva annua* var. *macrocarpa*), and sunflower (*Helianthus annuus* var. *macro-carpus*). In addition, the seeds of ragweed (*Ambrosia trifida*) stored with the known domesticates indicate that they, too, may have been har-vested from a garden and cached along with the other seeds, to be either planted or eaten the following spring.

Many, perhaps all, of the seeds buried there were never retrieved. The contents of the storage crevice burned, and when the crevice was excavated in 1934 all the seeds were carbonized (Henbest 1934). They are curated at the University of Arkansas Museum and constitute part of the data set examined in my doctoral dissertation (Fritz 1986). I re-consider them here in order to address several issues concerning the early stages of plant production in eastern North America: issues that can be discussed in greater depth today because of archaeobotanical advances made since 1986. The seeds from Marble Bluff are particularly useful in addressing the questions of when, where, and how quickly plant domestication occurred and, to a lesser degree, in assessing how important its products were for Terminal Archaic societies.

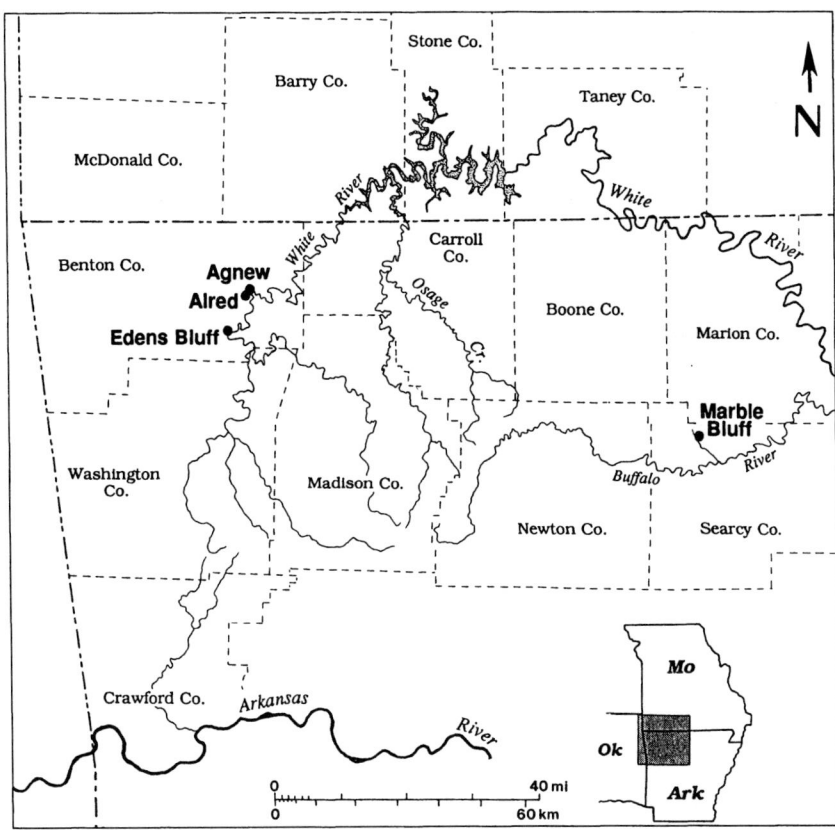

Figure 3.1. Map of the southern Ozark Highlands region, with locations of Marble Bluff and other sites.

Richard Yarnell has been, and still is, a key contributor and guiding force in the search for answers to these questions. His knowledge and advice were invaluable during my dissertation research, and I remain gratefully under his intellectual influence.

The Marble Bluff Site, 3Se1

The rockshelter known as Marble Bluff to the archaeological team from the University of Arkansas Museum is located in northwest Searcy County, Arkansas (figure 3.1), at an elevation of 245 m above mean sea level. The site is a long, narrow bluff overhang measuring approximately 120 m by 7 m (figure 3.2). At its lower, southern end the shelter

Figure 3.2. Photograph of Marble Bluff in 1990, facing south.

floor is only a few meters higher than Mill Creek, which flows directly beside the west-facing bluff overhang. The site is within the Spring-field Plateau subdivision of the Ozark Highlands, in the Buffalo River drainage. The Buffalo River itself is 5 km to the southeast of Marble Bluff. Terrain surrounding the site is rugged. The lowest elevation with-in a 1 km radius is 230 m and the highest is 380 m.

Excavations at Marble Bluff took place between March 30 and April 11, 1934, as part of an extensive program of investigations of Ozark rockshelter sites directed by Samuel C. Dellinger, Professor of Biology and Curator of the University of Arkansas Museum (Davis 1969). The excavations resulted in the cataloging of 405 samples includ-ing plant and animal remains, potsherds, stone and bone tools, cordage and basketry fragments, "buckskin," and either two or three human burials. Henbest's (1934) field notes record the depths from surface at which artifacts were encountered as ranging from 1 inch (2.5 cm) to 47 inches (119.4 cm). With the exception of carbonized specimens from the crevice, which I will discuss in this chapter, most plant remains from Marble Bluff were desiccated. The shelter was probably visited intermittently by prehistoric Indians in the area for at least 1,500 years, and probably more, judging by the presence of arrow points, potsherds, and corn (*Zea mays* ssp. *mays*), which was not a common crop in the Ozarks until approximately 1000–1100 B.P. (Fritz 1986).

The Cache Feature

The excavators found the crevice against the back wall of the shelter in excavation trench 63–64. At each end of the crevice "was a large rock causing a contained area for the cache. On top was a layer of shale, dust, and ashes" (Henbest 1934:100). The size of the area enclosed is not specified in the field notes, but Jerry Hilliard, Registrar of the Arkansas Archaeological Survey, examined the original site excavation map in the archives of the University of Arkansas Museum. Using grid lines (3 feet apart) for Block 3 and Trenches 62, 63, and 64, along with sketched-in outlines of the shelter's back wall and the large rock slab in front of the cache, Hilliard (personal communication 1993) estimated the maximum possible extent of the cache to be 7 feet by 3 feet. It seems unlikely that the dimensions exceeded 2 m by 1 m.

With the exception of one antler tool (MM-348), all artifacts and seed concentrations in the crevice are recorded as having been encountered at depths of either 16 inches (40.6 cm) or 17 inches (43.2 cm) below surface. The four pages of field notes documenting this feature include a sketch (figure 3.3), a list of the items recovered (summarized here in table 3.1), and a few sentences of general description ending with the following statement: "This whole cache was charred by fire after it was placed in this crevice between a slab and the wall." Field catalog numbers MM-326 through MM-348 (now 34-23-326 through 34-23-348) were assigned to the cache, with MM-326 evidently designating both the cache as a whole and a photograph of it curated by the University of Arkansas Museum (photograph negative number 340064).

Six catalog numbers were assigned to seed concentrations (table 3.1): five to bags of seeds and one to a deposit estimated by Henbest (1934) as having a volume of 1 gallon (3.8 l). Other items include three pointed wooden objects, five antler "objects," a notched bone "pendant," a polished quartz pebble, a "chip" of quartz crystal, two perforated mussel shells, a bag or net, and "basket pieces." To my knowledge, these objects have not been carefully examined. It is tempting to speculate that the pointed sticks and perforated mussel shells were cultivating tools. The bone and stone objects might also have played some role in the gardening process. If so, this cache could represent seeds stored for planting by a woman or group of women (or, conceivably, by a man or

Gayle J. Fritz

Figure 3.3. Redrawn sketch map of the Marble Bluff seed cache from page 98 of Wayne Henbest's 1934 field notes.

men on behalf of female gardeners) along with at least part of the tool kit needed to prepare the garden plot or plots.

The Age of the Seeds

Three samples of seeds from the Marble Bluff crevice were dated by the conventional (non–accelerator mass spectrometry [AMS]) radio-carbon method. An assay on ragweed achenes from sample 34-23-327, the 1-gallon pile of mixed seed types, resulted in an age determination of 2843 ± 44 B.P.:893 B.C. (SMU 1681) (Fritz 1986). The assay on chenopod seeds from the bag cataloged as sample 34-23-341 yielded a date of 2926 ± 40 B.P.:976 B.C. (SMU 1682) (Fritz 1986). Sumpweed achenes from the drawstring bag, sample 34-23-347, were dated to 2980 ± 30 B.P.:1030 B.C. (SMU 1874) (Fritz 1994). After dendrochronological cali-bration, using the CALIB program of Stuiver and Reimer (1993), the two-sigma (95% confidence interval) date ranges for the three samples are 1122–898 B.C., 1259–995 B.C., and 1301–1114 B.C., respectively. All three calibrated date ranges overlap at the 95% confidence interval, but

Table 3.1 Seed Bags and Other Artifacts from the Marble Bluff Crevice

Field Catalog Number	Description (Henbest 1934)	Trench Number	Depth (inches)
I. Seeds			
MM-327	"Seeds from cache 326 (1 gal)"	64	—
MM-340	"Bag with Seed" [chenopod]	64	16
MM-341	"Bag with Seed" [chenopod]	64	16
MM-344	"Bag of Seed" [chenopod]	64	17
MM-345	"Bag of large seeds & Basketry" [sunflower and squash]	63	17
MM-347	"Bag of large seeds with drawstring" [sumpweed]	63	16
II. Other			
.MM-326	"Cache Deposit" and "Photo"	64	16
MM-328	"Pointed Stick"	64	—
MM-329	"Pointed Stick"	64	16
MM-330	"Antler tool"	64	16
MM-331	"Antler tool"	64	16
MM-332	"Antler point"	63	16
MM-333	"Polished quartz pebble"	63	16
MM-334	"Pendant of bone with notches"	63	16
MM-335	"Antler object with hole in it"	63	16
MM-336	"Antler object with hole in it"	64	16
MM-337	"Mussel Shell with hole"	63	16
MM-338	"Mussel Shell with hole"	63	16
MM-339	"Crystal quartz (chip)"	63	16
MM-342	"Bag or net"	64	16
MM-343	"Basket pieces"	64	17
MM-346	"Pointed wooden object with the net 342"	63	16
MM-348	"Antler Tool"	64	12

the earliest and latest do not overlap at the 68% confidence level using the 1993 CALIB program: 1032–920 B.C., 1194–1028 B.C., and 1259–1128 B.C. A combined calibrated age range for the three dates is 1255–1241 B.C., 1216–1007 B.C. at the one-sigma level and 1264–912 B.C. at the two-sigma level (table 3.2).

Table 3.2 Radiocarbon Dates on Seeds from Marble Bluff

[14]C Lab Number	Material Submitted	Age (RCYBP)[1]	Calibrated One-Sigma Range[2]
SMU-1681	Ragweed from MM-327	2843 ± 44	1032–920 B.C.
SMU-1682	Chenopod from MM-341	2926 ± 40	1194–1028 B.C.
SMU-1874	Sumpweed from MM-347	2980 ± 30	1259–1128 B.C.

One-sigma combined calibrated age range: 1255–1241 B.C. 1216–1007 B.C.

Two-sigma combined calibrated age range: 1264–912 B.C..

[1]RCYBP = radiocarbon years before present.
[2]CALIB Rev. 3.0 (Stuiver and Reimer 1993).

The dated samples were recovered from different parts of the feature: the sumpweed bag was against the rear shelter wall in the northeast part of the crevice, the ragweed was near the center, and the chenopod bag was found near the southern end. Each of the three as yet undated bags from the cache would need to be assayed before all reservations about the contemporaneity of the six seed concentrations could be laid to rest, but available information supports the inference that all materials from this feature were deposited at the end of the fourth or beginning of the third millennium B.P. The scenario that makes most sense to me is that the cache represents one depositional episode: a single storage event. Alternative scenarios, of course, are plausible. Until evidence to the contrary is presented, however, I operate under the assumption that the six seed concentrations were deposited in the crevice at close to the same time and possibly on the same day.

Seeds from the Cache

Chenopod

Three of the five bags buried in the crevice at Marble Bluff contained thin-testa *Chenopodium berlandieri* ssp. *jonesianum* fruits. In addition, chenopod was the most abundant taxon by seed count in the concentration cataloged as sample 34-23-327. The seed bags have not survived excavation and curation intact. All have broken into smaller than fist-sized clumps of seeds that would not be recognizable as the

Crop Seeds from Marble Bluff

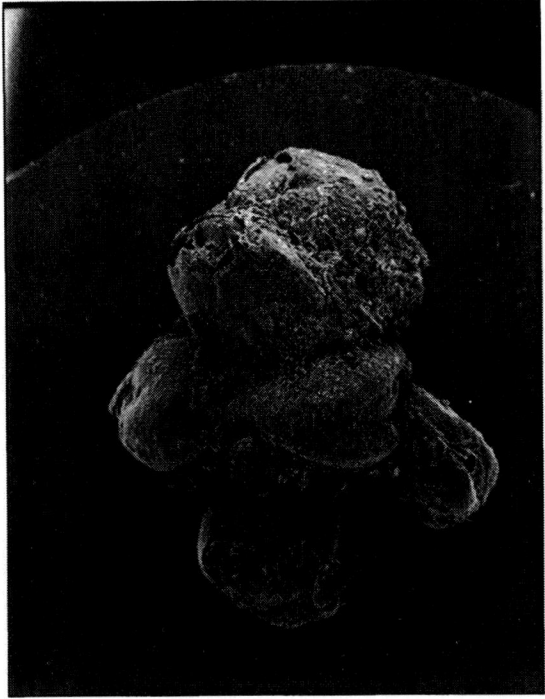

Figure 3.4. Scanning electron micrograph of a clump of carbonized chenopod seeds from sample 34-23-341.

contents of bags without close inspection. Some of these seed clumps have a convex or flat side consisting of twined cordage fragments from the original bag. One clump in sample 34-23-341 consists primarily of charred stems, probably grass stems. Remnants of cordage are stuck to the stems on one side of the clump, and a few chenopod seeds adhere to the cordage. This indicates that the storage pit was lined with grass and the bag full of chenopod seeds lay on the grass lining. Some seed clumps were evidently from the bags' interiors and include no recognizable cordage. It seems unlikely that the total content of any of these bags is in curation. Samples 34-23-340 and 34-23-344, in particular, are smaller than Henbest's sketch indicates.

The seed coats of chenopod specimens in sample 34-23-341 are thin (16 to 20 microns in thickness) and smooth. Shapes are difficult to discern because of puffing as a result of carbonization (figure 3.4), but they probably had truncate margins. Twenty-six seeds that had be-

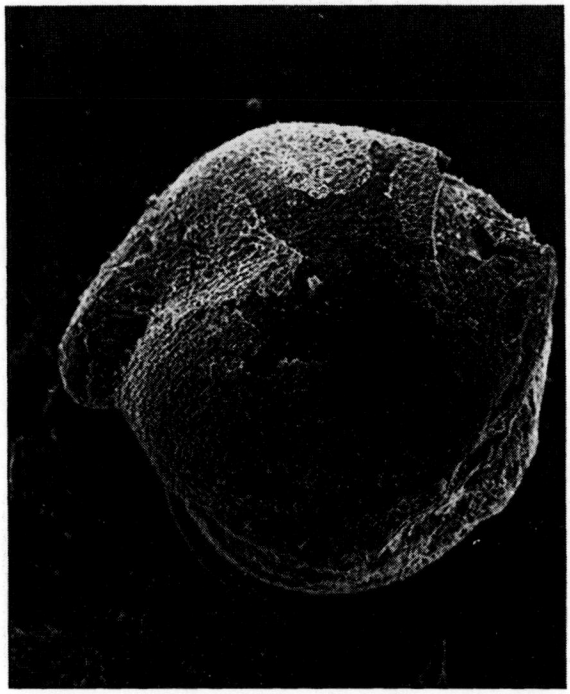

Figure 3.5. Scanning electron micrograph of a chenopod fruit from sample
34-23-327.

come detached from clumps in this sample were measured, and they
range in diameter from 1.0 mm to 1.7 mm, with a mean of 1.48 mm
and standard deviation of 0.17.

Although seeds from bags cataloged as 34-23-340 and 34-23-344
were not measured, preliminary inspection in 1984 led me to record
that they were morphologically like the seeds in 34-23-341 (Fritz
1986:84). This is not the case, however, for the chenopod seeds from
sample 34-23-327 (figure 3.5), the 1-gallon concentration of mixed
taxa from the central part of the crevice. Sample 34-23-327, as stated
above, was a pile of seeds that was probably originally contained in
some sort of net or basket, a remnant of which may have been collected as sample 34-23-343. Ragweed achenes appear to dominate this
sample in terms of mass, but they were outnumbered approximately
30:1 by the much smaller chenopod seeds in the subsample analyzed
(Fritz 1986:82). The seed concentration also includes, in descending or-

der of abundance by count, an unidentified taxon with four-seeded capsules suspected to belong to the mint family (Lamiaceae), sunflower, amaranth (*Amaranthus* sp.), knotweed (*Polygonum* sp., but probably not *Polygonum erectum*), gourd/squash, sumpweed, and maygrass (*Phalaris caroliniana*).

The chenopod specimens in this sample have thicker, more pitted seed coats than their counterparts in the three twined bags, and their margins are more rounded than truncate in shape (figure 3.5). Many are attached on one side and split open slightly on the other, resembling tiny clams. One seed coat measured using a scanning electron microscope was 25 microns thick, which is intermediate between the thickness of thin-testa cultigens (usually 12 to 20 microns) and that of wild forms (40 to 80 microns). Remnants of reticulate pericarp demonstrate that these specimens represent the species *C. berlandieri,* but the thicker, more pitted seed coats and rounded seed margins implicate them as possible weedy forms rather than clear domesticates or a fully wild population. Diameters of 200 chenopod seeds in the subsample ranged from 1.1 to 2.0 mm. The mean diameter of 1.60 mm (standard deviation 0.17) is slightly larger than that of the thinner-testa sample from bag 34-23-341.

This 1-gallon deposit of mixed seed types is obviously different from the five seed bags in the cache, each of which contains only one or two species of definitely domesticated plants. The more weedy-looking chenopod in the 1-gallon deposit, the presence of other weedy types such as amaranth, the inclusion of the unidentified Lamiaceae (?) and unusual knotweed, and the larger volume of this sample compared with that of the five seed bags argues for a distinct purpose for its storage. The bags of pure domesticates seem to be seed stock, intended for planting, whereas the 1-gallon deposit may represent the fruits of a generalized harvest intended, at least in part, for future consumption. Alternative scenarios, again, are plausible. Ragweed is a candidate for inclusion in Late Archaic and Early Woodland period food production systems (Asch and Asch 1985b; Fritz 1986), and it would not have been difficult to pick out achenes from the mixed deposit for sowing, had the gardeners so desired and the cache not burned.

Gourd/Squash Seeds

Cucurbita pepo seeds were present in two samples from the crevice: fragments of three were found in the analyzed subsample of 34-23-327

Gayle J. Fritz

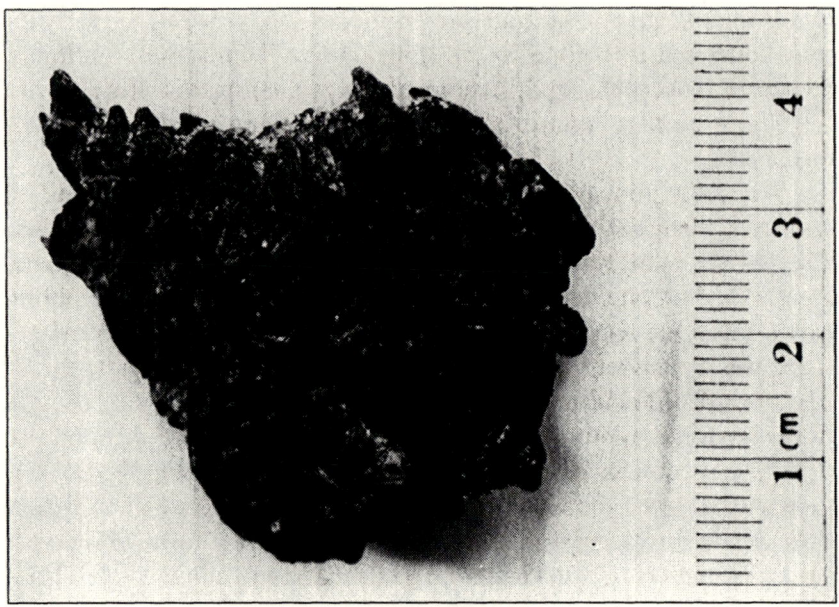

Figure 3.6. *Cucurbita* seeds and sunflower achenes from sample 34-23-345.

(the deposit of mixed taxa) and several dozen were found in a clump of sunflower and squash seeds constituting most of what remains of the bag with catalog number 34-23-345 (figure 3.6). I formerly felt confident about measurements for only four of the pepo seeds, whose entire outer surfaces were visible on the outside of the clump (Fritz 1986). After reexamination of the sample, however, I was able to obtain measurements for four more seeds (table 3.3).

The mean length of 11.6 mm is small for domesticated archaeological *Cucurbita* from the Ozarks. Only one other Ozark rockshelter, Agnew (3BE2), yielded a sample consisting of more than two measurable *C. pepo* seeds with an average length less than 11.6 mm. The thirty measurable seeds from Agnew sample 32-1-4 range in length from 9.9 to 11.6 mm and in width from 5.5 to 6.8 mm, with mean values of 10.8 and 6.3 mm, respectively. No radiocarbon assays were run on plant remains from Agnew, so the age of the sample is unknown. My guess would be that it is relatively early: older than 1500 B.P., but not necessarily older than the Marble Bluff sample despite its smaller average seed size. Sites near Agnew, such as Alred and Edens Bluff, were used by gardeners living along the White River during the Middle Wood-

Table 3.3 Measurements of *Cucurbita pepo*
Seeds in Sample 34–23–345

	Length (mm)	Width (mm)
	13.2	7.5
	12.7	7.8
	12.0	6.6
	11.8	8.5
	11.8	7.8
	11.3	—
	10.5	8.0
	9.7	7.0
Mean[1]	11.6	7.6
Standard deviation	1.06	0.59
Coefficient of variation	9.14	7.76

[1]Width/length ratio is 0.66.

land period (Fritz 1986), but no evidence is yet available for earlier seed storage in that area.

A great deal of progress has been made recently in charting the trajectory of gourd/squash domestication in eastern North America (Cowan et al. 1991; Cowan and Smith 1993; King 1985; Smith et al. 1992). Although debate continues, I agree with researchers who interpret the evidence as supporting independent domestication of a native gourd resembling *C. pepo* ssp. *pepo* var. *texana* rather than diffusion of an already domesticated tropical cultigen squash from Mesoamerica (Decker 1988; Heiser 1989; Smith 1989). Decker-Walters (Decker-Walters et al. 1993) has recently authored the epithet *C. pepo* ssp. *ovifera* var. *ozarkana* to designate modern free-living gourd populations in the Ozarks and elsewhere and, on the basis of allozyme analysis, suggests that this taxon was possibly ancestral to early cultigen *Cucurbita* gourds and squashes in the Eastern Woodlands.

Carbonized *Cucurbita pepo* rind fragments dating between 4500 and 7000 B.P. from Koster, Napoleon Hollow, and Carlston Annis are no thicker than the rind of modern free-growing gourds and likely reflect utilization of wild populations or plants in initial stages of a mutualistic relationship with people. The uncarbonized seeds and rind fragments recovered from Phillips Spring, Missouri, however, which date to ca. 4300 B.P., may well be products of domestication. The mean seed length for sixty-five measurable seeds is 10.5 mm, and seeds as large as 12.2 mm are present (King 1985:81). These values exceed both the mean and the maximum seed lengths of any free-living *Cucurbita pepo* population reported by Cowan et al. (1991) or King (1985). Decker and Wilson (1986) report one modern *C. pepo* gourd population from Illinois with a mean seed length of 10.5 mm. The experiences of David Asch (1994) and Leonard Blake (personal communication 1993) in Illinois and my personal conversations with farmers in Calhoun County, Illinois, however, make it prudent to reserve judgment about whether the majority of *C. pepo* gourds growing free in this state today are truly wild rather than naturalized from formerly cultivated fruits.

Moreover, *Cucurbita* seeds at Phillips Spring are associated with bottle gourd (*Lagenaria siceraria*) seeds and rind fragments that are more easily explained as cultigens than wild-growing plants. The 7,000-year-old bottle gourd at the Windover site in Florida may have been collected as a littoral-drift gourd (Doran et al. 1990), but it is difficult to envision that wild or feral bottle gourds were available in western Missouri during the Late Archaic period. Early cultivation of both *Cucurbita* and *Lagenaria* gourds at Phillips Spring seems more likely.

Smith (1987b) suggests an 11 mm mean seed length baseline value to separate wild from clearly domesticated *Cucurbita pepo* seed assemblages: using this baseline value "it is not until after 3,000 B.P. that average seed length values for *Cucurbita* exceed 11 mm (at Salts Cave and Cloudsplitter Rock Shelter in Kentucky)" (Smith 1987b:18). Even more emphatically, Smith (1992b:106) states: "Currently there is no evidence to support the proposition that a domesticated variety of *C. pepo* was present in the East prior to 3,000 B.P." In another recent paper, Smith et al. (1992:73) state, "the earliest evidence of clearly domesticated *C. pepo* in the region dates to about 3000 B.P.," but they do not specify the provenience.

The Marble Bluff *Cucurbita* seeds appear to provide the earliest clear morphological evidence of domestication for those uncertain about the Phillips Spring assemblage. The mean Marble Bluff seed length of 11.6 mm exceeds Smith's (1987b) suggested baseline value. Seeds from Marble Bluff are larger than those reported from Salts Cave paleofeces, which have a mean length of 11.3 mm and mean width of 7.3 mm (King 1985), in spite of the fact that the Marble Bluff seeds probably

Table 3.4 Size Information on the Marble Bluff Sunflower

Sample Number	Number Measured	Part Measured	Mean Achene L (mm)	Mean Achene W (mm)	L x W Index	W/L
34-23-327	13 L	Achenes	8.8		35	0.45
	15 W			4.0		
34-23-327	24 L	Kernels	9.3		39	0.45
	22 W			4.2		
34-23-345	19 L	Achenes	8.9		39	0.49
	18 W			4.4		
Higgs, Tennessee[1]	24	Unknown	7.8	3.1	24	0.40

Note: L = length; W = width. All dimensions are corrected for shrinkage during carbonization using Yarnell's (1978) correction factors.
[1]Data from Yarnell (1978).

experienced some shrinkage during carbonization. If sample 34-23-345 is accepted as coeval with the three directly dated Marble Bluff samples, which I argue is likely, then domestication of North American *Cucurbita pepo* was a fait accompli before 3000 B.P.

Sunflower

Sunflower achenes were mixed with *Cucurbita pepo* seeds in the clump from seed bag 34-23-345, discussed immediately above (figure 3.6), and both kernels (true seeds) and achenes (fruits) were present in the 1-gallon deposit of mixed seeds, sample 34-23-327. Specimens from the two samples are very similar in size and shape (table 3.4), a factor that strengthens the argument for their contemporaneity. The direct radiocarbon date on ragweed from the mixed seed deposit demonstrates its terminal Late Archaic period of origin. Because sunflower seeds shrink during carbonization, all measurements reported here have been "corrected" by multiplying kernel length by 1.30, kernel width by 1.45, achene length by 1.11, and achene width by 1.27 (following Yarnell 1978) to achieve an estimated precarbonized achene size. Sample 34-23-327 is divided into specimens that were measured as achenes and specimens that were measured as kernels, with the two categories yielding slightly different results. Adjusted mean achene length is 8.8 to 9.3 mm, and adjusted mean achene width is 4.0 to 4.4 mm.

Crites (1993) recently reported a direct, uncalibrated AMS date of

4265 ± 60 B.P. on a sample of sunflower achenes from central Tennessee with a mean length value of 6.9 mm. These specimens, from the Hayes site, range from 5.7 to 7.4 mm in length. Sunflower achenes longer than 7.0 mm are judged to have been harvested from domesticated plants (Heiser 1985). Marble Bluff and the Higgs site in eastern Tennessee share the distinction of yielding the next-earliest cultigen sunflower. The Higgs site sunflower, according to Yarnell (1978:296) is "securely dated to about 900 B.C." The specimens from Marble Bluff, however, are considerably larger (table 3.4). The index of mean length times width (L × W) (after correction for carbonization) for sunflower achenes from the Higgs site is 24 (Yarnell 1978), as compared with L × W values of 35 to 39 for achenes from Marble Bluff. One achene from a Middle Woodland context at the McFarland site in central Tennessee has reconstructed dimensions of 9.9 × 4.3 mm (Crites 1991). No other component predating the late second millennium B.P. is known to have yielded sunflower seeds or achenes as large as these from Marble Bluff (Yarnell 1978, 1981, personal communication 1993).

The unparalleled size, for their age, of the 3,000-year-old sunflower fruits from Marble Bluff is discussed below in connection with the similarly large sumpweed achenes in the same cache.

Sumpweed

One fragmentary sumpweed achene was found in the analyzed subsample from the mixed seed deposit, 34-23-327. Most sumpweed from Marble Bluff, however, came from sample 34-23-347, sketched and described as a drawstring bag full of large seeds (Henbest 1934). Five clumps of carbonized sumpweed achenes from this sample have been examined. The clumps weighed a total of 19.98 g, but some of the weight was contributed by fiber from the bag that held the seeds.

The lengths of eighteen achenes and widths of seventeen could be measured (table 3.5). Multiplication by a correction factor of 1.11 to adjust for shrinkage during carbonization yields a mean achene length of 6.5 mm (standard deviation 0.67) and mean width of 4.6 mm (standard deviation 0.48). The L × W index, accordingly, is 30. The age of this sample was determined by radiocarbon assay to be 2980 ± 30 B.P., as discussed above.

The sumpweed achenes are huge for their age, even more so than the Marble Bluff sunflower specimens. Whereas mean length for sunflower specimens from samples 34-23-327 and 34-23-345 is 1.0 to 1.5

Table 3.5 Size of Sumpweed Achenes from Marble Bluff Sample
34–23–347

	Number Measured	Range (mm)	Mean (mm)	Standard Deviation
Length	18	5.9–9.1	6.5	0.67
Width	17	4.3–6.2	4.6	0.48

Note: Length times width index = 30. All dimensions corrected for shrinkage during carbonization.

mm greater than that of their counterparts from other components predating 1500 B.P., mean length of sumpweed achenes from the drawstring bag is 2.0 to 3.0 mm longer than that of samples from Early Woodland period deposits in Salts Cave and Cold Oak shelter in Kentucky and more than 1.0 mm longer than mean lengths for Middle Woodland assemblages from west-central Illinois and central Tennessee (Asch and Asch 1985b; Crites 1985; Gremillion 1993c; Yarnell 1978). Not until ca. 1500 B.P. do sumpweed achene sizes surpass the ones from Marble Bluff, and these are other Ozark rockshelter samples (Fritz 1986). Outside the Ozarks, larger sumpweed achenes are Mississippian in age.

There are several possible reasons for this size difference:

1. The Marble Bluff samples, as carefully selected seed stock, represent only the extreme upper end of the size range.
2. The surviving Marble Bluff specimens did not shrink as much as loose seeds when charred because they were packed tightly and protected by other seeds, fiber bags, and pit lining.
3. The Marble Bluff specimens reflect one exceptionally favorable growing season and/or one group of exceptionally skillful gardeners.
4. Selection for larger sunflower and sumpweed seeds proceeded at a more rapid pace in the southern Ozark Highlands than elsewhere during the fourth millennium B.P.
5. The Marble Bluff gardeners acquired the seeds from another source area where selection had been more intense, or the Marble Bluff gardeners had recently moved into the Buffalo River drainage from this other place, bringing the large seeds with them.

Additional possibilities exist, no doubt, and most combinations of the above are not mutually exclusive.

Gayle J. Fritz

Four Issues in Early Eastern Agriculture: Date, Pace, Place, and Importance

Date and Pace

The 3,000-year-old cultigens from the Marble Bluff storage crevice are among the earliest specimens of eastern North American crops, but, with the exception of the *Cucurbita* seeds, none is a candidate for the earliest domesticate of its species. Thin-testa chenopod specimens from Cloudsplitter and Newt Kash rockshelters in Kentucky are some 500 years older than the ones from Marble Bluff (Smith and Cowan 1987). A larger-than-wild sumpweed assemblage from Napoleon Hollow in Illinois and sunflower from the Hayes site in central Tennessee have both been dated as more than 1,000 years older than the seeds from this crevice (Asch and Asch 1985b; Crites 1993).

Marble Bluff's *Cucurbita* seeds may be the earliest members of their taxon with mean length greater than Smith's (1987b) suggested 11 mm boundary, but 11 mm is a high baseline value for wild (not feral) *C. pepo.* I suspect the 4,300-year-old *C. pepo* seeds from Phillips Spring exhibit earlier and less obvious effects of domestication. As discussed above, the mean length (10.5 mm) and upper end of the Phillips Spring seed length range (12.2 mm) are higher than those of modern free-growing populations with the exception of one from Illinois that might well be naturalized rather than truly wild (Cowan et al. 1991; Decker and Wilson 1986; King 1985). The range of morphological variability in *Cucurbita* underscores the fact that although Smith's baseline values are heuristically useful, they are not solid boundary divisions. As Smith himself (1987b:40) points out, "it is not possible to establish a simple size boundary for distinguishing the seeds of domesticated vs. nondomesticated *C. pepo* taxa." Smith (1987b:40) quotes 1987 correspondence from Deena Decker to the effect that there "have been trends both towards smaller and larger seeds (as well as a lack of change in size) since initial domestication [so that] [w]hile large seeds can reasonably be assumed to represent domestication, it cannot be assumed that small seeds came from wild populations." The same appears to be true for *Cucurbita* rind thickness.

The Marble Bluff crevice deposit, then, is significant for containing four principal taxa of the so-called Eastern North American Agricultural Complex toward the end of the ca. 5000–3000 B.P. period during which they all exhibit morphological characteristics associated with domestication. Five bags and a probable bundle, the 1-gallon deposit, were placed in the same confined area for storage at the same depth. Three of the six samples, one from near the center and one from near each end of the crevice, have been directly dated and appear to be either exactly or approximately the same age. Morphological characteristics of seeds in the three undated samples give no cause for suspecting them of being younger than the others. I further speculate that older bags of nonviable seeds would have been removed from the pit had they been encountered there by people during a more recent storage event. Therefore I consider it likely that all of these samples are the same age.

The sunflower and sumpweed achenes are exceptionally large. Rather than falling along a curve of gradual size increase beginning earlier (ca. 5000–4000 B.P.) in the Late Archaic period, when domestication began, the Marble Bluff achenes are much larger than those in other known collections of Late Archaic and Early Woodland period sunflower and sumpweed samples. Even considering the circumstances of their unusual archaeological context, these samples indicate more rapid size increase than any previously available evidence. Selection pressures during the process of domestication may have been more intense in some regions than in others. Important evolutionary developments may have proceeded in a punctuated manner rather than at a gradual pace, at least in this part of the Eastern Woodlands.

Place

No localized hearth of initial domestication within the midlatitudinal, midcontinental Eastern Woodlands is recognizable. Smith's (1992a) floodplain weed theory of plant domestication highlights major river valleys where shoals, levees, and backwater sloughs supported rich aquatic resources that attracted human foragers. Other researchers (Fritz 1990; Gardner, this volume; Ison 1991; Munson 1986; Watson 1985) suggest the likelihood that primary domestication occurred either in more upland areas or along stream terraces, whether lowland or upland, where management of nut resources, rather than the pull of aquatic resources, triggered the sufficient combination of natural and cultural evolutionary mechanisms.

The Marble Bluff cache is not old enough to help resolve the issue in a direct manner. It does show, however, that plant husbandry was well developed as long ago as 3000–3300 B.P. in an upland zone far removed from backwater slough riverine environments. These people may have migrated into the Ozarks from the Arkansas or Mississippi river valleys, but so far no horticultural Late Archaic parent society is known in those regions.

Marble Bluff joins the list of sites with Archaic and Early Woodland period components where archaeobotanical remains shed light on the transition to food production in eastern North America. Unfortunately, none fits into a well-understood settlement system that can be monitored through time without major gaps from the Middle Archaic into the Middle Woodland period. Early use of domesticated plants can be documented in the lower Illinois River Valley (Asch and Asch 1985b), the Duck and lower Little Tennessee river valleys (Chapman and Shea 1981; Crites 1978, 1987, 1993), the Red River Gorge area (Cowan 1985b; Gremillion 1993c and this volume), the Green River Valley (Yarnell 1969, 1974a, 1974b; Watson 1985), the Pomme de Terre Valley (Chomko and Crawford 1978; Kay et al. 1980; King 1985), and here at a site on Mill Creek, 5 km from its juncture with the Buffalo River. Some site locations are in close proximity to expansive floodplain systems, some are on narrower, more upland terraces, and some, including Marble Bluff, have access to very little alluvial soil and few aquatic resources (see also summary in Watson 1985). Primary domestication may not have occurred independently in all of these settings, but none can be excluded from having contributed to the agricultural trajectory that Smith (1987b:37) eloquently describes as the "complex mosaic of occasionally linked, generally parallel, but distinct co-evolutionary histories for different areas of the mid-latitude eastern woodlands."

Economic Importance

The native Eastern seed crops—sunflower, sumpweed, chenopod, knotweed, maygrass, and little barley—have until recently been viewed by many archaeologists as generally insignificant in the prehistoric diet (e.g., Bender 1985; Fiedel 1987). Even today, with wider acknowledgment of their role as relatively important foods, the date at which they attained this status is uncertain. Smith (1989:1570) cites the Middle Woodland span of 250 B.C. to A.D. 200 (2200–1750 B.P.) as being the time of "initial and additive emergence of multicrop food-producing economies in eastern North America," which I take to mean subsistence systems in which domesticated plants played a major role. Evidence pointing to their importance during the Late Archaic period is absent from most regions. Cowan (1985b) and Gremillion (1993c and this volume), for example, present evidence for heavier use of crop seeds during the Early Woodland than during the preceding Late or Terminal Archaic period in eastern Kentucky.

Assessment of the dietary significance of specific food types is problematic to unwise even under the best controlled of archaeological circumstances (Fritz 1994; Hastorf 1988; Pearsall 1989), and the problems are compounded when one deals with rockshelter assemblages (see also the discussion by Gremillion, this volume). Most plant specimens from Marble Bluff are uncarbonized. The charred crevice cache is a unique situation, and it may hold the only Terminal Archaic archaeobotanical remains recovered from the site. Crop seed types in the charred bags are not represented in desiccated samples from noncrevice contexts. Available radiocarbon dates are limited to the three assays from the crevice, but a much longer span of site use is indicated by the presence of maize, which could date as early as ca. 2000 B.P. or as late as historic times. It most likely postdates 1500 B.P., however, as do the arrow points illustrated by Henbest (1934).

Some of the uncarbonized plant types—nuts, seeds, and fruits—may have been brought to the shelter by rodents or by larger nonhuman mammals. Sampling practices during the 1930s were less systematic than we would expect them to be today, and the 1934 field notes are not complete enough for assessment of the integrity of many contexts. Finally, my analysis focused on domesticated and cultivated plants. Samples in which these taxa were absent have not been fully sorted. Therefore a quantified evaluation of the significance of domesticated plants in the overall subsistence system as reflected at Marble Bluff is not possible at this time.

In spite of these limitations, I conclude with the admittedly and purposefully vague suggestion that the Marble Bluff seed cache attests to the importance of the food-producing segment of the economy of this Terminal Archaic society. The crops were probably far less important at 3000 B.P. than they were 1,000 years later. It is unlikely that they exceeded 20% to 30% of the caloric intake of the people who grew them, and the average proportion per capita may well have been lower. However, gardening was clearly beyond the experimental stage. A good deal of effort was put into the processing and storage of these seeds. A weedy chenopod may have evolved and established a niche for itself in the plots where domesticated chenopod was planted. The sunflower, sumpweed, and gourd/squash seeds were large and carefully selected. Their availability, especially at lean times of the year, may have reduced mortality, increased fertility, reduced the size of the necessary catchment zone, enhanced social interactions, and spurred movements of groups (or segments budding off from them) into territories previously unoccupied or occupied only by nonhorticulturists. As Smith (1989:1568) summarizes, early (pre-2500 B.P.) gardens probably "played a significant role in providing a dependable, managed, and storable food supply for late winter to early spring."

An unusual and fortuitous set of circumstances led to the preservation and recovery of the seeds in the Marble Bluff crevice. Although they constitute a rare survival of Terminal Archaic crop seed storage, they reflect a typical subset of activities in the general process of early food production. Yarnell recognized the significance of these activities at least twenty-five years ago, when he reported on the high proportion of sumpweed, sunflower, chenopod, gourd/squash, and other probable crop seeds in the Salts Cave paleofeces (Yarnell 1969, 1974b). The Marble Bluff cache, small but loaded with valuable material, serves as a well-developed forerunner of more intensive agricultural manifestations. The fruits of the labors of the gardeners who buried these seeds in Marble Bluff were consumed by fire, but subsequent generations, including ours, benefited from their work.

Acknowledgments

To the staff of the University Museum, University of Arkansas at Fayetteville, I again extend gratitude for assistance along with respect for effective curation. I would like to thank Jerry Hilliard for estimating the dimensions of the Marble Bluff cache and Jim Railey for drawing both the regional map and the reproduction of the sketch map of the cache. Kristen Gremillion, Lee Newsom, Bruce Smith, Patty Jo Watson, Richard Yarnell, and one anonymous reviewer provided valuable editorial comments. This work would have been impossible without Dick Yarnell's pioneering research into eastern North American plant domestication and his patient mentoring. A shorter version of this chapter was presented at the 50th Southeastern Archaeological Conference, Raleigh, in November 1993.

CHAPTER 4

Evolutionary Changes Associated with the Domestication of *Cucurbita pepo*

Evidence from Eastern Kentucky

C. WESLEY COWAN

Thanks to the "flotation revolution," the broad outline of the evolution of field agriculture in eastern North America is arguably the best such record in the world (Cowan and Watson 1992; Smith 1992). The past two decades, for example, have seen the formal identification of two extinct domesticates (sumpweed [*Iva annua* var. *macrocarpa*] [Yarnell 1972] and chenopod [*Chenopodium berlandieri* ssp. *jonesianum*] [Smith and Funk 1985]) as well as three other weedy annuals that were intensively cultivated in various portions of the Midcontinent (maygrass [*Phalaris caroliniana*], erect knotweed [*Polygonum erectum*], and little barley [*Hordeum pusillum*] [Asch and Asch 1978; Cowan 1978a; Smith 1992b]). Along with the familiar sunflower (*Helianthus annuus*), which was also probably domesticated by eastern populations, these plants form a distinctive crop group whose evolution occurred in the North American Midwest and Southeast sometime between about 6000 and 3000 B.P. It is safe to suggest that at least among eastern archaeologists and many plant taxonomists this developmental trajectory is widely accepted.

In addition to these relatively obscure—at least by public standards—annuals, there is another plant whose status as an eastern domesticate has been the subject of intense debate in recent years. "Squash," or at least a primitive form of *Cucurbita pepo*,[1] has also been found in archaeological and paleontological deposits in widely spaced areas in the eastern United States in contexts that predate the intensive use of the weedy annuals. This debate focuses on the issue of whether *Cucurbita*

C. Wesley Cowan

pepo was introduced into eastern North America from a Mesoamerican homeland or was domesticated in the East independently of Mexico (see Asch and Asch 1992; Cowan and Smith 1993; Decker-Walters et al. 1993; Smith 1992; and Smith et al. 1992 for recent summaries of the various positions).

Recent genetic evidence suggests that at least one lineage of *Cucurbita pepo* (ssp. *ovifera*) evolved in situ in eastern North America (Decker-Walters et al. 1993). The progenitor of this lineage seems to have been similar to, if not identical to, wild gourds whose modern distribution encompasses a broad area of the southeastern United States (Cowan and Smith 1993; Smith et al. 1992). The occurrence of abundant *C. pepo* seeds in a paleontological site in Florida directly dated to the thirteenth millennium B.P. adds considerable strength to the argument that a wild gourd has been a member of an indigenous eastern flora since at least the end of the Pleistocene (Newsom et al. 1993). These parallel lines of evidence point to the independent domestication of at least some varieties of summer squashes and ornamental gourds (those placed in *Cucurbita pepo* ssp. *ovifera* var. *ovifera*) within eastern North America.

In spite of the fact that a likely progenitor for these important crops has been identified, knowledge of the evolution of *Cucurbita pepo* spp. *ovifera* in the East remains incomplete. In part, our understanding of this process is hindered by physical and cultural factors that place limits on the kind and size of archaeological remains available for study. *Cucurbita* remains are generally not common in any archaeological contexts, but particularly so in eastern North America where varying soil types and microorganisms, temperature, humidity, and moisture fluctuation interact to limit the preservation of all types of plant parts. Carbonization is the primary process by which most archaeobotanical assemblages are preserved, and those plants with lignified cell structures stand a better chance to withstand flame than those with softer tissue. These anatomical features favor nutshell and large seeds with hard coats over tubers and wood over leaves, for example.

The process of carbonization is particularly unkind to cucurbits. Both squash and *Cucurbita* gourd rinds are composed largely of loosely organized rows of parenchyma cells (which are soft, thin-walled, and easily warped and eroded) and outer layers of harder, thicker-walled sclerenchyma cells. Both are easily consumed by heat. Excavation and flotation of sediments further degrades carbonized archaeological cucurbit specimens. The combination of these factors more often than not reduces archaeological *Cucurbita* rinds to small blackened bits that are not of much use beyond simple identification. Other parts of the plant (e.g., seeds, peduncles, rind warts, and blossom or peduncle scars) are even less frequently recovered or recognizable.

Fortunately, cucurbits are also occasionally preserved in uncarbonized states in dry cave or rockshelter deposits or in perpetually waterlogged sediments. In these situations, preservation can be quite spectacular (e.g., Gilmore 1931: plate XXIV; Watson 1969: plate 12B) and cucurbit parts can be abundant. Remains recovered from these sorts of depositional environments are of great value in helping us understand the evolution of the genus *Cucurbita* because they retain characteristics generally missing from their charred counterparts. Distortion and loss from carbonization is reduced, more parts of the plant may be preserved, and even original coloration of the fragile epidermis may be retained.

This chapter summarizes *Cucurbita* remains from perpetually arid rockshelters in the rugged Cumberland Plateau region of eastern Kentucky and adds to the growing literature of the history and evolution of one group of the family Cucurbitaceae: *Cucurbita pepo* ssp. *ovifera*. The collections from these overhangs provide dramatic evidence of the initial use of a wild *Cucurbita* by as early as 5100 B.P., followed by its domestication more than two millennia later. This evolutionary process was accompanied by changes in rind characters and seed and peduncle size, which document the transformation of a presumably inedible *Cucurbita* gourd to edible forms of squash and provide a window on the emergence of food production as an element of the premaize economies of the midcontinent.

Cucurbits from the Cumberland Plateau

Cucurbita rinds, peduncles, and seeds representing both wild and domesticated forms have been recovered from a number of rockshelters in the Licking, Red, and Kentucky river drainages of the Cumberland Plateau in Kentucky. Their dates extend from those of Archaic contexts as old as 5100 B.P. to those of Late Woodland deposits as recent as 1200 B.P. (figure 4.1, table 4.1). The research potential of these materials varies considerably from site to site and largely reflects the development of interest in the recovery of archaeobotanical materials during the past two decades (see Gremillion, this volume).

Many of the shelters in Wolfe, Powell, Menifee, and Lee counties

C. Wesley Cowan

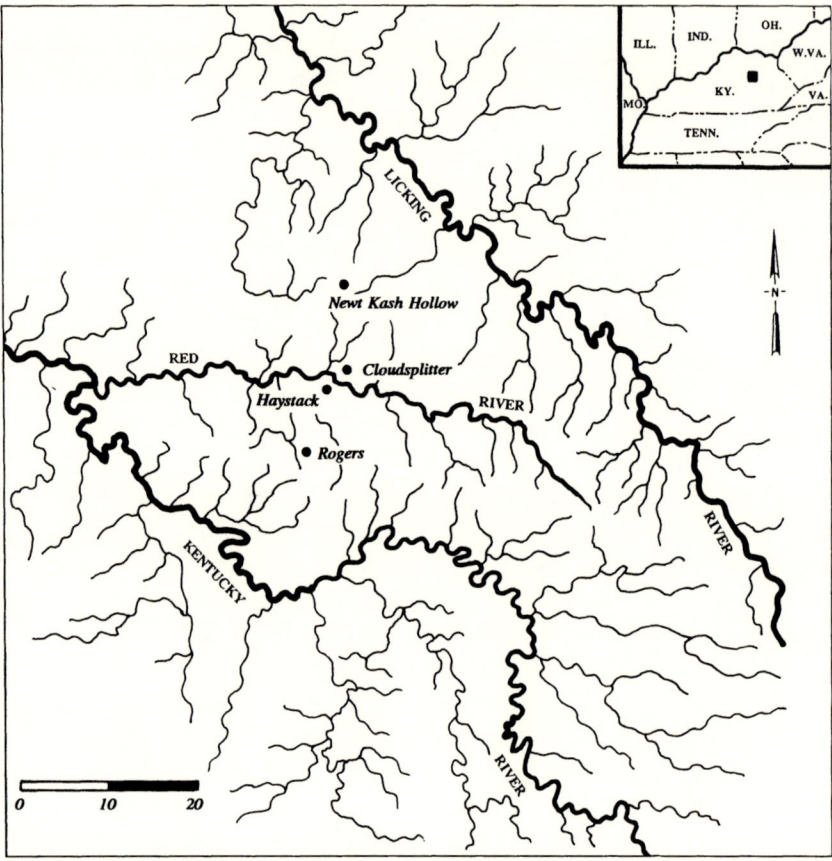

Figure 4.1. Location of the Cumberland Plateau in eastern Kentucky and important sites discussed in the text. Scale in kilometers.

excavated by the University of Kentucky during the late 1920s and early 1930s were reported to have produced *Cucurbita* and other plant remains (Funkhouser and Webb 1929, 1930; Haag 1974; Webb and Funkhouser 1936). Botanical materials were collected systematically, however, at only two sites, the Newt Kash (Webb and Funkhouser 1936) and Hooton Hollow (Haag 1974) shelters. Of these, the Newt Kash remains are best known through the pioneering work of the late Volney Jones (Jones 1936). As part of this study the Hooton Hollow remains were examined, but were found to be unsuitable for analysis (see below).

More recent excavations of eastern Kentucky shelters designed explicitly to recover plant remains have yielded abundant *Cucurbita* re-

Table 4.1 Rockshelter Sites Producing Cucurbit Remains in the
Cumberland Plateau

Site Name	County	Drainage	Age[1]	References
Newt Kash Hollow	Menifee	Licking	E.W.	Webb and Funkhouser 1936
Hooton Hollow	Menifee	Red	E.W.	Haag 1974
Cloudsplitter	Menifee	Red	L.A.–E.W.	Cowan et al. 1981; Cowan 1985a
Rogers	Powell	Red	L.W.	Cowan 1979b
Haystack	Powell	Red	L.W.	Cowan 1975, 1979a
DeHart	Powell	Red	?	Funkhouser and Webb 1930
Buckner	Lee	Kentucky	?	Funkhouser and Webb 1929
Cold Oak	Lee	Kentucky	L.A.	Gremillion 1993c and personal communication

[1]L.A. = Late Archaic; E.W. = Early Woodland.

mains from Archaic, Early Woodland, and Late Woodland contexts (cf.
Cowan 1978a, 1979a, 1979b, 1985a; Cowan et al. 1981; Gremillion
1993c). Each of these sites shares a single defining characteristic that led
to the preservation of normally perishable cucurbit remains. The depths
or ceiling heights of the overhangs might vary, but they are sufficient
to protect the interior from atmospheric precipitation. As a consequence,
the overhang floor is basically a moistureless desert that acts to preserve
organic materials, both cultural and biological, in a desiccated form. In
essence, whatever organic material blew or fell into the overhang or was
brought to the overhang by human or other occupants stood an excel-
lent chance of being preserved by the natural aridity of the interior.

Archaic Cucurbits

Archaic-aged (ca. 5000–3000 B.P.) *Cucurbita* gourds have been recovered from deposits at two Cumberland Plateau overhangs, the
Cloudsplitter and Cold Oak shelters (figure 4.1, table 4.1). At Cloudsplitter, the Archaic sample consists of eleven rind fragments and two

C. Wesley Cowan

Table 4.2. Radiocarbon Dates Associated with Cucurbits from Eastern Kentucky

Site	Sample	Radiocarbon Age	Calibrated RCYBP[1]
Cloudsplitter[2]	Beta–46700	5130 ± 60	3960 B.C.*
	Ariz–3358	4700 ± 250	3500, 3420, 3380 B.C.*
	UCLA–2313k	3620 ± 80	1960 B.C.
	UCLA–2313j	3550 ± 60	1880 B.C.
	UCLA–2340n	3370 ± 100	1670 B.C.
	Beta–46701	3150 ± 55	1410 B.C.*
	UCLA–2313h	3060 ± 60	1310 B.C.
	UCLA–2313a	2710 ± 60	830 B.C.
	UCLA–2313f	2710 ± 60	830 B.C.
	UCLA–2313c	2615 ± 60	800 B.C.
	UCLA–2340c	2440 ± 80	510, 430 B.C.
	UCLA–2313d	2370 ± 60	400 B.C.
Cold Oak	Beta–38051	2900 ± 100	1040 B.C.*
Newt Kash	M–31a	2650 ± 300	810 B.C.
	M–31b	2600 ± 300	800 B.C.
Rogers	UGa–552	1485 ± 55	A.D. 600
	UGa–553	1470 ± 65	A.D. 610
	UGa–553	1415 ± 60	A.D. 650
	UGa–749	1345 ± 60	A.D. 670
	UGa–750	1245 ± 60	A.D. 780
Haystack	GX–5875	1405 ± 120	A.D. 650

[1]Calibrated using University of Washington Quaternary Isotope Lab Radiocarbon Calibration Program CALIB Rev. 3.0 (Stuiver and Reimer 1993). RCYBP = radiocarbon years before present.
[2]Radiometric dates from Cloudsplitter bracket deposits containing *Cucurbita* remains. Those marked by an asterisk represent *Cucurbita* rind or seed directly dated through AMS technique.

seeds from eight proveniences. Two rinds and one seed have yielded direct accelerator mass spectrometer radiocarbon (AMS) dates, while the remainder were recovered from deposits bracketed by a number of traditional radiometric dates (Cowan et al. 1981: tables 1 and 2; table 4.2, this chapter). These determinations indicate that a *Cucurbita* gourd first makes its appearance in the Cloudsplitter record as early as 5100 B.P. and occurs thereafter throughout Late Archaic deposits dated between 4700 and 3000 B.P.

The rind fragments from Cloudsplitter are all from small, thin-walled fruits with a smooth epidermal layer. None is large enough to determine the shape or exact size of the fruits (but see below). The thickness of nine of the eleven rind fragments is measurable. Only two of the nine rinds exceed 2 mm in thickness, and both of these occur stratigraphically above rinds less than 2 mm in thickness (table 4.3). If these two thicker rinds are excluded from the sample, the mean thickness of the Late Archaic fruits is 1.5 mm. Two of these rinds have yielded direct AMS dates of 5130 ± 60 B.P. and 3150 ± 55 B.P. (table 4.2). At Cold Oak, rinds ranging in thickness from 2.6 to 3.5 mm are associated with Terminal Archaic activity at the site; one of these specimens has been directly dated to 2900 ± 100 B.P. (tables 4.2 and 4.3).

In addition to the rind fragments from Cloudsplitter, two small seeds were also recovered with mean dimensions of 8.7 mm (length) and 5.4 mm (width) (table 4.4); a portion of one seed has yielded an AMS date of 4700 ± 250 B.P. (table 4.2). From a comparison of the Cloudsplitter seeds with those of modern wild forms of *C. pepo* ssp. *ovifera* var. *ozarkana* and *C. pepo* spp. *ovifera* var. *texana,* as well as a number of cultivated varieties (i.e., domesticated forms of *C. pepo* spp. *ovifera*), the Archaic Kentucky gourds were small, with fruits probably averaging less than 10 cm in height and about 5 cm in diameter and weighing no more than a few grams (Cowan and Smith 1993; King 1985: figure 4.8).

The morphological characteristics of the rinds and seeds from Cloudsplitter are strikingly similar to those of both the Ozark gourd (Cowan and Smith 1993; Decker-Walters et al. 1993), whose distribution includes riparian habitats throughout the highlands of the Arkansas and Missouri Ozarks (Cowan and Smith 1993; Smith et al. 1992), and the Texas gourd (Decker 1988) of the central part of that state. A similar gourd has been present in Florida at least since the end of the Pleistocene (Newsom et al. 1993). Presumably, at least some of the *Cucurbita* specimens that have been recorded in several Archaic-aged contexts in the Southeast and Midwest (Asch and Asch 1992; Crites 1991; Kay et al. 1980) originated from a similar wild gourd. These occurrences suggest that a *Cucurbita pepo* gourd whose prehistoric distribution may have included the Cumberland Plateau of Kentucky is an ancient member of an indigenous eastern flora (Cowan and Smith 1993).

Table 4.3. *Cucurbita* Rind Thickness for Cumberland Plateau Rockshelter Sites[1]

| | Fruit Type | | |
	Smooth (n) Range Mean, S.D.[2]	Lobed (n) Range Mean, S.D.	Warty (n) Range Mean, S.D.
Cloudsplitter Archaic	(9) 1.00–2.40 1.66, 0.36		
Cold Oak Terminal Archaic/ Early Woodland	(1) 2.6		(2) 3.0–3.5
Cloudsplitter Early Woodland	(16) 1.00–2.70 1.91, 0.47	(10) 1.60–3.00 2.25, 0.42	(3) 1.80–3.00 2.33, 0.61
Newt Kash Early Woodland	(48) 0.60–4.10 2.49, 0.75	(29) 1.40–4.20 2.76, 0.68	(2) 2.90–3.00 2.95, 0.07
Rogers Late Woodland	(401) 0.50–6.30 1.37, 0.54	(113) 1.00–3.60 1.80, 0.63	(67) 1.00–3.00 1.94, 0.57

[1]All measurements in millimeters.
[2]S.D. = standard deviation.

None of the Archaic rind fragments shows any evidence of having been a portion of a container. Indeed, on the basis of a comparison of the eastern Kentucky archaeological specimens with wild Ozark and Texas gourds, it is unlikely that Archaic foragers would have found them to have much utility as vessels, rattles, or floats, all of which have been proposed as uses for early gourds (Ford 1981; Prentice 1986). Wild *Cucurbita* gourds are too small and their shells too brittle to have effectively served any of these proposed uses. While the utility of wild gourd shells as containers can be questioned, the potential of the seeds as sources of food cannot. A preliminary study of the seeds of the Ozark gourd found

Table 4.4. *Cucurbita* Seed Dimensions from Eastern Kentucky Rockshelters[1]

Site	N	Length Range Mean, S.D.[2]	Width Range Mean, S.D.
Cloudsplitter Archaic	2	8.6–8.8 8.7, 0.14	5.2–5.6 5.4, 0.28
Cloudsplitter Early Woodland	5	10.6–13.7 12.6, 1.2	6.2–7.9 7.4, 0.71
Newt Kash Hollow Early Woodland	2	12.1–14.2 13.1, 1.45	8.6–9.7 9.1, 0.77
Haystack Late Woodland	3	11.0–12.0 11.3, 0.57	6.0–7.0 6.3, 0.55
Rogers Late Woodland	363	9.0–13.7 11.6, 1.0	6.0–9.6 7.3, 0.56

[1]All measurements in millimeters.
[2]S.D. = standard deviation.

them to be good sources of dietary fat and protein (Smith 1992; see also Adams 1975:133 for domesticated varieties of squashes and pumpkins). It is this character that probably initially attracted Archaic foragers and served as the impetus to cultivate wild gourds. Their visibility within the natural world and the ease by which a local population can be manipulated make wild gourds ideal candidates for the sort of experimentation and casual cultivation that many believe characterized early horticulture in eastern North America.

Wild *Cucurbita* gourds are denizens of the floodplain. Where present along sand and gravel bars or in flotsam deposited by spring floods, they are highly visible and impressive producers (Cowan and Smith 1993; Smith et al. 1992). The size of a wild gourd patch could have been easily increased by simply scattering seeds from flood-deposited fruits in their natural habitats.

The Context of Archaic Cucurbit Cultivation

Like their counterparts elsewhere in eastern North America, preagricultural foragers in the Cumberland Plateau were highly dependent on the seasonal rhythm of production of plant and animal resources. While a few mammalian species were the focus of hunting and the source for most preferred foods, plants formed the mainstay of the subsistence economy, with nuts a key fall season resource. Interannual variability of nut production, however, placed a ceiling on the degree to which mast was a dependable food source for foragers. In years when a "bumper" mast crop occurred, surplus production could be cached away as a form of "insurance policy" that could be drawn on in a lean season. When production was simply average, or worse, below average, foragers faced extreme deprivation in the late winter and early spring months. Since most nut bearers in the East do not begin production until the trees are about thirty years old, there is little human beings can do to manipulate nut crops effectively on a short-term basis.

It was in this context that the protein-rich seeds of wild *Cucurbita* gourds (and the other weedy annuals that were eventually domesticated or intensively cultivated) came into play. Because gourd production is so easy to manipulate, expanding the area in which gourds grow could produce a sizable crop. More important, the harvest of seeds from this sort of incipient "garden" could function as yet another sort of "insurance policy," providing food to buffer against the interannual variability of nuts. This sort of "garden of gourds" (cf. Bailey 1937) is envisioned as one of the first conscious alterations of a plant population undertaken by human foragers in the Cumberland Plateau.

Early Woodland Cucurbits

Three rockshelters (Cloudsplitter, Newt Kash, and Hooton Hollow) have yielded more than 100 *Cucurbita* rind fragments and seeds assignable to the Early Woodland (3000–1500 B.P.) period. The discussion that follows is based primarily on the Cloudsplitter and Newt Kash remains. While being curated at the University of Kentucky in the 1930s, the *Cucurbita* remains from Hooton Hollow were heavily cleaned and abraded, resulting in considerable attrition of the innermost layers of cells of the rind fragments. Although useful for determining morphological characteristics (i.e., fruit form), the collection is not suitable for metric analysis (i.e., measuring rind thickness) and for this reason is excluded from the present study.

Early Woodland *Cucurbita*-bearing deposits at Cloudsplitter are bracketed by numerous radiocarbon dates ranging from about 2800 to 2300 B.P. (table 4.2). No Early Woodland rinds or seeds have been directly dated. The Newt Kash collection is essentially "floating" in the sense that it lacks adequate provenience information from within the overhang. Artifacts recovered from the site are predominantly Terminal Archaic/Early Woodland in age, although a few projectile points were recovered from both earlier and later occupations (cf. Webb and Funkhouser 1936: fig. 16). Radiometric determinations partially confirm these affiliations. A recent AMS date on five seeds of domesticated goosefoot (*Chenopodium berlandieri* ssp. *jonesianum*), for example, yielded a corrected age of 3400 ± 150 B.P. (Smith and Cowan 1987) and a sample of paleofeces produced an AMS date of 3025 ± 55 B.P. (Gremillion 1994). Two Newt Kash samples dated in the 1950s by the University of Michigan's Phoenix Memorial Laboratory produced solid carbon dates of 2600 ± 300 and 2650 ± 300 B.P. (Crane 1956). From the artifact assemblage from the site and corroborating radiometric determinations, the Newt Kash *Cucurbita* remains described here are assumed to have been deposited on the overhang floor during the Terminal Archaic and Early Woodland periods, with most relating to the Woodland occupation of the site.

The Early Woodland *Cucurbita* rinds and seeds from Cloudsplitter and Newt Kash are quite different from their Archaic counterparts. Both fruit morphology and metric data indicate that the wild *Cucurbita* gourd utilized/cultivated by Archaic foragers underwent a significant transformation between 5000 and 3000 B.P. (tables 4.2 and 4.3). These changes mark the development of a domesticated variety of *Cucurbita pepo* spp. *ovifera* and the emergence of forms of primitive, but almost certainly edible, squash. In the Cumberland Plateau this evolution is evidenced by (1) a diversification of fruit forms and accompanying epidermal lobing and wartiness, (2) an increase in both the size of fruits and the seeds they contained, and (3) a marked change in the thickness of the fruit wall or rind.

From the evidence of rind characteristics, three fruit morphotypes—a smooth-shelled form, a lobed form (some lobed rinds also have widely scattered warts), and a warty form—were grown by the Early Woodland inhabitants of both Cloudsplitter and Newt Kash. A warty variety was also utilized as early as ca. 3000 B.P. at the Cold Oak shelter. None of the fragments is large enough to permit estimates of fruit size, but judging by the curvature of the rind sherds, fruits were small and similar in shape to several modern varieties of ornamental gourds and the 'Mandan' cultivar of *C. pepo* ssp. *ovifera* var. *ovifera*. Similarly aged nearly complete mature fruits from Salts Cave in central Kentucky average about 15 cm in diameter (range 12–25 cm) (Wilson n.d.; Yarnell 1969:51). Even though the size of the eastern Kentucky fruits cannot be accurately predicted on the basis of the rind fragments collected from Cloudsplitter or Newt Kash, it is clear they were larger than the Late Archaic *C. pepo* gourds from the same sites.

Changing fruit morphology was also accompanied by a dramatic increase in seed size, offering further support for an increase in overall fruit size. The five Early Woodland–aged *Cucurbita* seeds found at Cloudsplitter are 30% larger in mean length and 37% larger in mean width than their Archaic counterparts. Two seeds from presumably Early Woodland contexts at Newt Kash are even larger (table 4.4). Seeds of similar size from contemporary *Cucurbita pepo* cultivars produce mature fruits weighing between 1 and 1.5 kg (2–3 pounds) (see King 1985: fig. 4.8).

The epidermal characters exhibited by the Early Woodland populations mark important deviations from the ancestral wild gourd. Lobing and wartiness are not present in any eastern North American or Mexican wild gourd population that has been described (Andres 1987; Cowan and Smith 1993). Such traits, however, are common in both ornamental and edible modern cultivars of *C. pepo* ssp. *ovifera*. These morphological changes, along with an increase in overall size of fruits and seeds, point to an Early Woodland divergence of a domesticated *C. pepo* from some wild ancestral stock.

Epidermal warts and lobing were also accompanied by an increase in rind thickness and, presumably, hardness. While there is some degree of overlap, the three morphotypes of Early Woodland *Cucurbita* gourds and squashes can be sorted on the basis of rind thickness (table 4.3). In general, smooth fruits had the thinnest rinds and warty fruits the thickest; lobed fruits were intermediary in thickness. These differences are hardly observable for the combined populations of Cloudsplitter and

Newt Kash, but when the two populations are separated, an interesting pattern emerges (table 4.3). The Cloudsplitter rind fragments are thinner than those from Newt Kash, suggesting a possible developmental sequence of thin-walled to thicker-walled fruits. This proposed sequential relationship needs to be confirmed, however, through AMS dating of both Cloudsplitter and Newt Kash *Cucurbita* rinds.

The Context of Early Woodland Cucurbit Cultivation

Rind hardness also provides clues about gardening strategies, the timing of utilization of the Cumberland Plateau shelters, and the context of *Cucurbita* cultivation within the Early Woodland subsistence economy. By and large, fruits of contemporary edible cultivars of *C. pepo* ssp. *ovifera,* the so-called summer squashes, are eaten in an immature state. If not picked and eaten, the flesh of many of these cultivars becomes increasingly fibrous and less palatable as the growing season progresses. In their fully mature state some are hardly edible, and the shell, or rind, becomes sclerified, or "woody." Allowed to mature and dry, even the thin wall of a yellow crookneck is capable of retaining the original shape of the fruit, although it is hardly durable. While it is impossible to know whether Early Woodland squash fruits were eaten in their immature form, the hard, "woody" rinds recovered from the eastern Kentucky shelters do provide significant evidence that at least a portion of the crop was allowed to mature fully. Mature fruits also would have yielded two products not associated with their immature counterparts: large, edible seeds and hard shells.

If both immature and mature fruits were eaten, this implies a complex human settlement pattern in the Cumberland Plateau that currently is only poorly understood. The earliest date Late Archaic/Early Woodland gardeners could safely plant a crop without fear of killing frost lies between the first and second weeks of May. Immature but edible fruits would have been available about 60 days later, by the middle or end of July. Completely mature fruits would have been ready for harvest beginning in the latter part of August and continuing until the first killing frost, sometime late in October. Early Woodland rockshelter utilization, however, seems to have centered on the fall and winter months. At Cloudsplitter, large thermal features associated with interior structures and deep subfloor pits characterize what was probably a winter occupation. A midspring presence is suggested by a small, rock-lined storage cist that may have held the seed stock for an Early Woodland garden (Cowan 1985a). There is no evidence of summer use of the overhang.

If Cloudsplitter and other Early Woodland period overhangs were primarily used in the fall and winter, at least a short spring occupancy must be inferred from the presence of crop plants that dictated an early to mid–May planting schedule in order to mature. Since Cloudsplitter was not occupied during the summer or early fall, residents must have left the overhang at this time of year to pursue other activities before returning in the late fall.

By the Early Woodland period, cucurbits were only one component of a complex gardening regime that included domesticated sumpweed, chenopod, sunflower, and bottle gourd (*Lagenaria siceraria*), as well as maygrass and erect knotweed, two plants whose status as domesticates remains unclear. Produce from cultivated plants may have formed a significant component of the diet of eastern Kentucky foragers by this time (see Gremillion 1994), but still functioned to even out periods of shortage caused by variation in mast production.

Late Woodland Cucurbits

Cucurbita pepo remains have also been recovered from the Late Woodland Rogers and Haystack shelters (Cowan 1975, 1979a, 1979b). Both sites have been radiocarbon dated to between approximately 1300 and 1400 B.P. (table 4.2). Gourd/squash remains from Haystack are limited to three small rind fragments of fruits with a smooth epidermis and three seeds (Cowan 1979a:9). More than 1,000 *C. pepo* rind fragments, seeds, and peduncles were recovered through excavations at Rogers, however, making this one of the largest and best-preserved populations of archaeological *Cucurbita* from the Midcontinent.

The size and condition of the Rogers *Cucurbita* collection permit fairly detailed descriptions of Late Woodland fruits (Cowan 1979b:143–53). These collections indicate that (1) Late Woodland gardeners continued to grow at least three forms of *C. pepo* fruits, (2) post-1500 B.P. fruits of *C. pepo* were perhaps no larger than Early Woodland fruits, but they were more fleshy, and (3) Late Woodland fruits are similar to nineteenth- and early twentieth-century varieties of *Cucurbita pepo* ssp. *ovifera* squash grown historically by American Indian farmers along the middle Missouri River.

There is little indication that the forms of fruits grown by Late Woodland populations differed from Early Woodland fruit forms since each of the three kinds identified at Newt Kash and Cloudsplitter is present in the Rogers collection. Late Woodland rinds of all types, however, are significantly thinner than those recovered from Early Woodland contexts (table 4.3). In fact, mean rind thickness of the most common type (smooth shelled) is thinner than that of the Archaic counterparts from presumably wild *Cucurbita* gourds. The dramatic reduction in rind thickness that is represented in the Late Woodland collections suggests that the squashes grown by A.D. 600 were bred primarily for fleshy edible mesocarp as opposed to edible seeds or a durable shell that could be used as a container.

Most of the rinds from Rogers appear to have originated from smooth-shelled fruits, though many are quite small and could well be fragments of other fruit types.[2] In fact, because the smooth-shelled fruits also possess the thinnest rinds, they probably have a tendency to break into more and smaller pieces than the thicker-walled forms. Judging from the largest rind pieces, smooth-shelled fruits were variable in size and often pyriform in shape. One fragment represents a fruit at least 15 cm in diameter. Of this type, fourteen smooth rind portions retaining the peduncle scar and one fragment retaining the blossom scar were recovered.

Larger fragments of ribbed or lobed forms indicate both a completely smooth form and a slightly to strongly warty form were being grown. Fruits are invariably ten lobed, a common characteristic in domesticated *Cucurbita pepo* (Tapley et al. 1937). Several fragments indicate that the blossom end of the fruit had a prominent "knob" or raised blossom scar. One fragment from the peduncle end of a fruit is bulbous, suggesting a pyriform shape. Eight lobed fragments retaining the blossom scar and four retaining the peduncle scar were recovered.

Warts on fruits range from scattered to densely packed. Individual warts range from 2 to nearly 16 mm in diameter and are raised from 0.5 to 7 mm above the surface of the rind. Some of the warty fruits are strongly to shallowly lobed. Two peduncle end fragments and one blossom end fragment of this fruit form are included in the collection.

A total of seventy-eight individual peduncles were recovered from the Rogers middens. All are weakly to strongly five angled and thus easily distinguishable from bottle gourd (*Lagenaria siceraria*) peduncles. Diameters of the peduncles at their point of attachment with the squash

C. Wesley Cowan

Table 4.5. Peduncle Diameters for Various Wild and Domesticated *Cucurbita* Populations

Collection	Species	N	Mean	Range	S.D.
Rogers	*C. pepo*	78	11.6	4.0–15.4	2.1
Ozarks[1]	var. *ozarkana*	20	6.6	5.3–8.6	0.8
Ohio[2]	ssp. *ovifera*	23	9.7	6.2–15.1	2.8

S.D. = standard deviation.
[1]From Cowan and Smith 1993: table 5.
[2]From a sample of gourds cultivated in Hamilton County, Ohio, reported by Cowan and Smith 1993: table 5.

fruit range from 4 to 15.4 mm, with a mean of 11.6 mm (standard deviation 2.17). These are typical of peduncle diameters for contemporary varieties of summer squashes and for many ornamental gourds (*Cucurbita pepo* spp. *ovifera* var. *ovifera*). They are distinctly larger, however, than those of the wild Ozark gourd (var. *ozarkana*) and of some contemporary populations of var. *ovifera* (table 4.5; Cowan and Smith 1993). Not surprisingly, the range of peduncle diameters reveals considerable overlap between the Rogers *C. pepo* and contemporary varieties of ssp. *ovifera*.

Some of the Rogers shelter rinds retain residual traces of their epidermal coloring. By far the majority range from light ivory to reddish brown, but others are clearly green striped, green, or yellow. Similar coloration is present in many of the modern ornamental var. *ovifera* gourds and *Cucurbita pepo* cultivars. Ivory and green-striped colors also characterize the Ozark gourds (Cowan and Smith 1993).

In addition to the rinds and peduncles, more than 400 complete and fragmentary squash seeds were recovered from Rogers. Seed length of 363 complete seeds ranges from 9.0 to 13.7 mm, with a mean of 11.6 mm (standard deviation 1.0). Widths range from 6.0 to 9.6 mm, with a mean of 7.3 (standard deviation 0.56). These mean dimensions are approximately 25% larger than those of the Archaic seeds from Cloudsplitter and fit comfortably within the range of variability expressed by the small Early Woodland sample. The relatively low coefficients of variation (CV= 100 × standard deviation/sample mean) for seed length (8.62) and width (7.67) for the population, however, mask the apparent differences in fruit morphology represented by the rind fragments and imply that at least for this collection, seed morphology is not useful for determining fruit shape within *C. pepo*.

The Context of Late Woodland Cucurbit Cultivation

Late Woodland rockshelter utilization in the Cumberland Plateau suggests a pattern of seasonal use similar to that of both Terminal Archaic and Early Woodland populations. At both Rogers and Haystack seasonal indicators imply primarily a late fall occupation. Lacking are the large storage cists, ash dumps, and prepared hearths that characterize what were probably winter occupations by Early Woodland peoples.

Human coprolites recovered from Haystack contain large quantities of bark and plant cuticle, which may represent foodstuffs consumed during the spring of the year (Cowan 1978b). Cultivated plants are virtually absent from these coprolites, in stark contrast to the contents of some of the Terminal Archaic feces from Newt Kash and Hooton Hollow recently analyzed by Gremillion (1994). The Late Woodland pattern is consistent with a regular winter abandonment of the Cumberland Plateau, followed by a spring visit to plant a garden that could be left and then harvested during an annual fall visit. These visits were initiated from large, nucleated agricultural villages that may have witnessed intensive nearly year-round occupations (Oetelaar 1990; Wymer 1990; see also below). In this regard, the Cumberland Plateau gardens simply reflect the continued use of cultivation as an "insurance policy" to counteract the local unpredictability of nut production. It should be stressed, however, that the eastern Kentucky gardens were strictly local responses; the same family groups who seasonally migrated to the Cumberland Plateau may have been engaged in intensive field agriculture near their permanent villages (Wymer 1993).

Evolutionary Trends

The fortuitous preservation of plant remains in dry rockshelters of eastern Kentucky provides a discontinuous but nearly 5,000-year window on the utilization and evolution of *Cucurbita pepo* gourds and squashes in the Midcontinent. While unique from the standpoint of preservation, the rinds, seeds, and peduncles from the Cumberland Plateau of eastern Kentucky record a pattern of development that probably took place elsewhere throughout the Mississippi River drainage and perhaps beyond.

The Cumberland Plateau record is one that began with the use of a wild *Cucurbita* gourd similar to, if not identical to, the Ozark (*C. pepo* ssp. *ovifera* var. *ozarkana*) or Texas (*C. pepo* ssp. *ovifera* var. *texana*) gourds that grow today throughout much of the lower Mississippi basin (Cowan and Smith 1993; Decker-Walters et al. 1993; Smith et al. 1992). While adventive *Cucurbita* species have been reported growing in the Cumberland Plateau, no truly wild gourd is part of a contemporary flora of this region. If the evolutionary sequence presented here is correct, a wild *Cucurbita* gourd probably enjoyed a much wider range in prehistory than it does today, and that range may have included the western edge of the Cumberland Plateau in eastern Kentucky (Decker-Walters 1990; Heiser 1985; Smith 1989).

If the wild *Cucurbita* gourds that were collected by Late Archaic populations in eastern Kentucky were like their modern counterparts in the Ozarks and elsewhere in the Southeast, they were small and thin-shelled with bitter flesh, but contained abundant, protein-rich seeds (see Cowan and Smith 1993 for a discussion of the characteristics of wild Ozark gourds). It was this characteristic, coupled with the local abundance of the wild gourds, that attracted human use. Initially, it was their value as sources of food—emphatically *not* as containers—that led foragers to harvest and probably cultivate wild gourds.

The data reported here indicate that between 4000 and 2500 B.P. the wild *Cucurbita pepo* gourds used by Late Archaic populations underwent significant morphological evolution. Rind thickness increased (figure 4.2), and epidermal characteristics of the eastern Kentucky archaeological collections imply that three fruit morphotypes were being grown by 2800–2500 B.P.: a smooth-shelled, round to pyriform form similar to today's "egg gourd," a round to oblate warty form similar to present fancy varieties of *C. pepo* ssp. *ovifera* var. *ovifera,* and a lobed and sometimes warty form closest in form to the historic "Mandan" variety grown by agricultural groups occupying the middle reaches of the Missouri River.

Unlike their Late Archaic ancestors, at least two of the Early Woodland varieties were thick-walled and large enough to have been utilized as durable, lightweight containers. Terminal Archaic material from Cold Oak dating to ca. 3000 B.P. (Gremillion 1993c) may represent the initiation of this morphological trend. Although no vessel fragments were identified in the more than 100 eastern Kentucky rind fragments examined, similarly aged *C. pepo* fruits recovered in Salts Cave in west-

Domestication of *Cucurbita pepo*, Eastern Kentucky

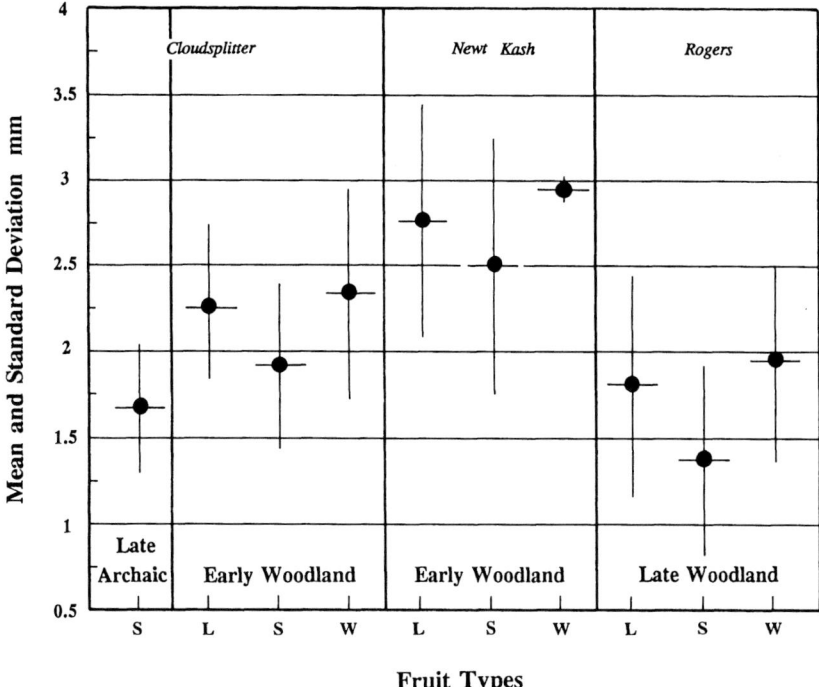

Figure 4.2. Changing rind thickness in eastern Kentucky *Cucurbita* collections. S = smooth or nontextured fruit type; L = lobed fruit; W = warty fruit.

central Kentucky provide unequivocal evidence that large fruits were occasionally used for vessels (Wilson n.d.; Yarnell 1969:51–52).

Gross size of seeds also changed tremendously during this 1,500 year interval. Although the sample of Archaic and Early Woodland *C. pepo* seeds is regrettably small, the few that were recovered indicate that size increased by more than a third during the period between 4000 and 2500 B.P. Seed size and fruit weight are highly intercorrelated in both wild and cultivated *Cucurbita;* large seeds are produced by large fruits (cf. King 1985: fig. 4.8). The dimensional changes in seed size and rind thickness documented in the 4000 and 2500 B.P. eastern Kentucky collections probably reflect the emergence of domesticated cultivars of *Cucurbita pepo* spp. *ovifera*. AMS dating of individual rinds and seeds may help to narrow this two and a half millennium window of domestication.

The number of *Cucurbita* remains recovered from Archaic and Early

Woodland contexts in the Cumberland Plateau is small: 117 rind fragments and 9 seeds. This may reflect an early, low-level use of *Cucurbita* between 5000 and 3000 B.P. and sharply contrasts with the incidence of remains in later sites. By the latter part of the second millennium B.P. remains of *C. pepo* are ubiquitous in the Late Woodland Newtown phase deposits at Rogers: more than 1,500 squash rinds, seeds, and peduncles were collected from the partial excavation of this small site.

The growing importance of edible squashes is reflected not only locally, but also regionally by the sheer quantity of *Cucurbita pepo* remains recovered from archaeological contexts. At Rogers alone, that quantity represents a tenfold increase over the combined number from Early Woodland shelters in the Cumberland Plateau. A similar pattern of abundance is also characteristic of coeval open-air Late Woodland sites elsewhere in the Ohio Valley. At the Zencor/Scioto Trails site in the Middle Scioto drainage of Central Ohio, for example, carbonized rind of *C. pepo* occurred in 82% of the twenty-three flotation samples analyzed and in 39% of the twenty-three samples not analyzed by flotation (Wymer 1993). At Childers, a Newtown or Childers phase village on the banks of the Ohio River in Mason County, West Virginia, rinds occurred in more than 80% of the 101 samples analyzed (Wymer 1990). When considered in this context, the Late Woodland eastern Kentucky rockshelter material probably reflects a regional trend toward increased cultivation of *Cucurbita pepo* ssp. *ovifera* for food.

Not only do *Cucurbita* remains become more abundant in Late Woodland contexts, but they also reveal a change in morphology between 2500 and 1500 B.P. that probably reflects the importance of *Cucurbita* as a source of food. On the average, Late Woodland rinds are significantly thinner than their Early Woodland counterparts. Smooth-shelled fruit walls, for example, are reduced by 60% compared with Early Woodland material of the same type; lobed fruit walls are reduced by almost 40% and those of warty fruits by 36% (figure 4.2). This reduction in rind thickness almost certainly reflects an increasing effort to breed thick-fleshed, highly edible fruits. If *Cucurbita* was bred for rind thickness during the Early Woodland period, this characteristic seems to have been of little consequence to Late Woodland populations.

Late Woodland squashes are most similar to some of the *Cucurbita pepo* cultivars grown by American Indians in the Middle Missouri River drainage (Gilmore 1977; Tapley et al. 1937). Historically, these fruits were eaten in their immature stage or, if mature, were cut into long spirals or "doughnuts" and dried for future use.

In the absence of a Middle Woodland archaeological record from the Cumberland Plateau, it is only possible to suggest that the change from hard to fleshy fruits that took place in *Cucurbita pepo* cultivars must have taken place sometime in the millennium between about 2500 and 1500 B.P. This, of course, assumes a steady decrease in rind thickness and a general increase in both fruit and seed size. As fruits and seeds became larger, flesh became thicker, and hard-walled *Cucurbita* gourds were transformed into palatable forms of squashes that are recognizable today.

Conclusions

Desiccated *Cucurbita* remains from eastern Kentucky rockshelters provide a rare opportunity to document in detail the poorly understood evolutionary history of *Cucurbita pepo* in the Midcontinent. From the data presented here, the earliest form of *Cucurbita pepo* used by prehistoric eastern Kentuckians was a wild gourd with an exceedingly thin and brittle shell and probably bitter flesh. Although locally extinct in the Cumberland Plateau of Kentucky, a similar species continues to thrive today in the greater lower Mississippi drainage. Recent fieldwork in the Southeast (e.g., Cowan and Smith 1993; Decker-Walters et al. 1993; Smith et al. 1992) suggests the ancestral eastern Kentucky species was the Ozark gourd (*C. pepo* ssp. *ovifera* var. *ozarkana*) or a related lineal ancestor (e.g., *C. pepo* ssp. *ovifera* var. *texana*).

The sequence of eastern Kentucky archaeological populations reveals that significant morphological changes in the wild *Cucurbita* gourd occurred fairly rapidly between 4000 and 3000 B.P. Fruit size, seed size, and rind thickness all increased during this interval, reflecting selection for larger and presumably fleshier fruits. Fruit walls reached their maximum degree of sclerification (as measured in rind thickness), and larger specimens would have made durable lightweight containers. Perhaps not coincidentally, these evolutionary changes take place during the same interval when the first ceramic vessels begin to appear in Terminal Archaic/Early Woodland contexts.

By 1500 B.P., *Cucurbita pepo* remains from eastern Kentucky Late Woodland shelters indicate that fruits with thick, hard rinds were no longer the focus of gardeners. Late Woodland *C. pepo* rind is exceedingly thin compared with chronologically earlier material, reflecting an emphasis on breeding for meaty fruits. Presumably, the emergence of fruits with more edible flesh occurred sometime during the 1,500-year interval between 3000 and 1500 B.P.

This evolution occurred within variable cultural contexts. Archaic foragers may have begun the process as early as 5000 B.P. by simply spreading wild gourd seeds along sand banks and other favorable habitats. This simple cultivation did little other than to increase the numbers of harvestable gourd fruits. The emergence of distinct Early Woodland fruit morphotypes by about 3000 B.P. implies the careful selection and retention of useful random mutations that probably began during the Terminal Archaic. This selection process may go hand in hand with a pattern of increasing residential permanency that is evident in Terminal Archaic and Early Woodland settlement patterns in the Cumberland Plateau (Cowan 1985a; Gremillion 1993c, 1994). Neither Archaic nor Early Woodland rockshelter occupations were permanent, however. Both suggest a pattern of seasonal abandonment that included late fall and winter use, followed by spring planting and then abandonment for the summer months.

By about 3000 B.P. cucurbits were only one type of a number of cultivated plants grown in Early Woodland gardens. This suite included the bottle gourd, domesticated sunflower, sumpweed, goosefoot, and the quasidomesticates maygrass and erect knotweed. The dietary contribution these plants made to the Early Woodland peoples is not readily quantifiable, but functioned to even out the unpredictability inherent in a subsistence economy focused on nuts as a storable winter food commodity (Cowan 1985a).

Late Woodland residency of the Plateau was restricted to short-term visits in the fall, perhaps to hunt selected animal prey species, followed by winter abandonment. Late Woodland populations seem to have spent the winter months in large nucleated settlements located in regions outside the Cumberland Plateau. During the midspring, small family groups probably dispersed from these villages, some of which returned to Plateau rockshelters in order to plant small gardens that could be harvested during a return trip in the fall. After the gardens were planted, the families went to the larger settlement, spending the summer months harvesting aquatic resources and mammalian resources, before removing to the Plateau for fall hunting.

Acknowledgments

Many of the cucurbits discussed in this paper are curated at the Museums of Anthropology at the Universities of Kentucky and Michigan. Thanks are due to Richard I. Ford for access to the Newt Kash Hollow and Cloudsplitter materials curated in Ann Arbor and to Mary Lucas Powell for access to the Newt Kash Hollow, Haystack, and Hooton Hollow materials curated in Lexington. The materials from the Rogers shelters are curated at the Red River Historical Society Museum in Clay City, Kentucky, and I am indebted to Larry Meadows for allowing me to reexamine the squashes and gourds after so many years. Kris Gremillion generously provided unpublished *Cucurbita pepo* rind measurements and an associated direct date from the Cold Oak shelter. Funds for excavations and analysis of these materials were received from the National Park Service, the National Science Foundation, and the Cincinnati Museum of Natural History. The present version of this chapter benefited from the comments of Lee Newsom, one anonymous reviewer, and my cucurbit-loving colleagues Bruce D. Smith and Patty Jo Watson. Al Adamson drafted the skillfully executed line drawings in figures 4.1 and 4.2.

Notes

1. This chapter follows current nomenclature for the genus *Cucurbita* (Decker-Walters et al. 1993). This system recognizes two distinct lineages of *Cucurbita pepo*. One of these, *Cucurbita pepo* ssp. *pepo*, comprises the true pumpkins, marrows, Mexican landraces, and a few ornamental gourds. The other, *Cucurbita pepo* ssp. *ovifera*, includes the scallop and crookneck squashes and most domesticated ornamental gourds (*C. pepo* ssp. *ovifera* var. *ovifera*) and two wild *C. pepo* gourds (the Texas gourd [*C. pepo* ssp. *ovifera* var. *texana*] and the Ozark gourd [*C. pepo* ssp. *ovifera* var. *ozarkana*]). This lineage appears to be native to eastern North America (Decker-Walters et al. 1993). "Gourd/squash," "*Cucurbita* gourd," and "*C. pepo* gourd" as used here all refer to *C. pepo* ssp. *ovifera*. The term *squash* is applied only to this taxon's edible, fleshy cultivars.

2. In addition to rind fragments that could be assigned to one of three fruit types, the Rogers collection also contains fifty-one fragments of peduncle scars and forty-four blossom scars from unclassifiable fruits.

CHAPTER 5

Anthropogenesis in Prehistoric Northeastern Japan

GARY W. CRAWFORD

Paleoethnobotany in Japan has a relatively young history, especially with respect to intensive flotation sampling and interpretation of resulting data in the context of culture historical, processual, and other issues. Today, nearly fifty sites from a variety of periods have been examined by a handful of researchers using flotation. Recovery of plant remains from wet sites, which are relatively common in Japan, has also been emphasized in recent years. Although important data on cultigens are often derived from this work, they tend to be interpreted rather loosely, that is, without reference to taphonomic and depositional issues. In contrast, my own research has concentrated on recovering carbonized plant remains from dry sites in northeastern Japan and investigating them using a rigorous analytic framework.

This chapter examines prehistoric anthropogenesis in northeastern Japan, an issue that has become increasingly well defined during the past two decades because of an ever-improving comparative data base. After first setting out a brief overview of the importance of anthropogenesis, I review perspectives on this issue in Japan. I present data from sites in northeastern Japan that serve to illustrate the human impact on the environment there, using weeds as an indicator of anthropogenesis for the sake of this discussion. The carbonized seed data are extensive and span a period from about 8000 B.P. to 1000 B.P. Subsistence regimes range from foraging in the earliest periods to relatively substantial agriculture by 1200–1000 B.P. I consider patterns consistent with four phases of subsistence and anthropogenesis: Initial Jomon (9500–7500 B.P.), Early through Late Jomon (7500–3000 B.P.), Zoku-Jomon (2300–1600 B.P.), and Yayoi-Ezo (after 1600 B.P.).

Anthropogenesis

Anthropogenesis is the process by which human beings impact their environment. The resulting effects are manifested in nonequilibrium ecological states characterized by spatial and temporal patchiness (Reice 1994). Disturbance is the main factor involved in nonequilibrium ecological states. By removing organisms such as trees from a habitat, human beings set in motion processes that change the character of the ecosystem. This disruption, if relatively severe, takes the system back to less mature successional stages. An extreme example is monocrop agriculture; in contrast, a single tree-fall exemplifies a minimal, localized effect on an ecosystem. Young successional stages are characterized by rapid reproductive rates, short life cycles, and a high ratio of production to respiration, resulting in high net production and high gross production in relation to standing biomass (Odum 1971). These characteristics mean that greater quantities of fruit and vegetal materials, as well as greater numbers of animals such as deer and rodents, thrive in disrupted ecosystems. Thus ecological disruption can be detrimental in some forms, of course, but the advantages to human beings, at least in the short term, are obvious. Anthropogenesis is a critical factor in the success of human cultures. Ecological disturbance as a normal part of ecosystems must also come to be acknowledged in environmental policy development (Reice 1994).

Evidence for anthropogenic impact may be detected in paleosols, pollen profiles, wood charcoal assemblages, carbonized seeds, and the like. Of course, one must distinguish intentional activities such as field preparation from processes that produce similar effects, although not primarily intended to create new plant communities (e.g., extraction of construction materials, village periphery and interior disturbance). McCorriston and Hole (1991), for example, argue for the role of anthropogenesis in conjunction with other factors in the origin of agriculture in southwestern Asia. They argue that each region where agriculture began should be elucidated in its own terms (1991:10) as do many of the authors in Gebauer and Price (1992). In similar fashion, this chapter examines anthropogenic conditions correlating with the development of food production in prehistoric Japan. In addition, anthropogenesis contributes to our understanding of the remarkable success of the Jomon in northeastern Japan. I examine the evidence for human impact on vegetation, components of which became part of human subsistence. The body of data is carbonized plant remains from northeastern Japanese occupation sites spanning some 7,000 years. I pay special attention to herbaceous weeds, particularly annuals that colonize disturbed habitats, produce large numbers of seeds, and have rapid growth and high phenotypic plasticity.

Yarnell (1963) and others have argued for the importance of the concept of reciprocity in human ecology. This perspective is not common in Japanese research, which generally views people as relatively passive recipients of the naturally rich resources of the archipelago. Human impact on the environment is also seen as a unidirectional process of environmental modification to facilitate agriculture. However, a process of adjustment to changes implemented unintentionally, as well as intentionally, should be examined. Nearly a decade ago, I found evidence for Jomon people taking advantage of what appeared to be anthropogenic communities (Crawford 1983). This study expands upon my earlier discussion.

Prehistory and Vegetation History: An Overview

The culture history of Holocene northeastern Japan is complex. The view of a linear progression from earliest Jomon foragers to the Ainu, their presumed descendants in Tohoku and Hokkaido, is no longer accepted. Ainu culture is an outcome of several thousand years of northeast-southwest interaction, the results of which include the development of extensive food production throughout most of the northeast by 1000 B.P. (Crawford 1992; Crawford and Takamiya 1992; Crawford and Yoshizaki 1987; Howell 1994). This culture history sits within a backdrop of relatively well-defined vegetation and climate history. Most of the environmental history has been interpreted from pollen evidence, although other types of information, such as mollusk distribution, contribute to regional environmental reconstruction. These studies tend to view human beings as passive players on the prehistoric Japanese stage, at least until the advent of rice paddy agriculture in southwestern Japan about 2400 B.P.

From 8500 to 3500 B.P., a warm period is associated with the development of deciduous hardwood forests in the northeast and broadleaf evergreen forests in the southwest (Yasuda 1978). Agriculturally induced changes in vegetation are continuous in southwestern Japan starting about 3200 B.P. (Tsukada et al. 1986). There is some evidence for earlier forest clearance, for example, in the Ubuka Bog area, by 7700 B.P. (Tsukada et al. 1986) and at the Torihama Shell Mound, where extensive forest destruction is argued to have occurred about the same time (Yasuda 1978). Although preceded by these early impacts, the advent of rice agriculture left an indelible mark on the Japanese landscape. Beginning in the Yayoi (ca. 2400 B.P.), floodplain and hillside forests were extensively cleared (Yasuda 1978:242). Pine trees increased in abundance at the same time, apparently because of successional processes. These vegetation history interpretations tend to view forest clearance for agriculture as the primary detectable form of anthropogenesis. Regional pollen profiles are not normally able to elucidate the local effects of human communities. As a result, pre-Yayoi anthropogenesis is difficult to assess from regional pollen profiles. Fortunately, many projects routinely include the analysis of pollen from habitation sites. Site-specific pollen records such as from the Yagi site, for example, indicate that pollens of grass and knotweed (*Polygonaceae*), among other herbaceous weedy groups, have a significantly higher influx rate than those of arboreal taxa (Davis 1979).

Nishida (1980, 1981, 1983) examined nut remains and wood charcoal from wet sites in southwestern Japan. He proposed that Jomon people created orchards of nut trees near sites such as Torihama and Kuwagaishimo. The nut cultivation model has gained popularity in recent years and is another version of the contention that Jomon plant food was comprised mainly of nuts. This is an oversimplification that, while acknowledging the potential role people played in affecting the character of local habitats, does not address the broader range of environmental interactions during the Holocene in Japan.

Weed Occurrence Patterns in Sites in Northeastern Japan

At least four types of plant remains assemblages seem to correspond with general cultural developments in northeastern Japan. The first appears in a relatively small set of samples taken from Initial Jomon contexts at Nakano B in Hakodate (figures 5.1 and 5.2). The second is associated with the subsequent Early through Late Jomon phases (figure 5.2). The Zoku-Jomon is a third pattern (figure 5.3). The fourth is associated with the agriculturally oriented Yayoi and Ezo/Heian of Aomori and Hokkaido (figures 5.4 and 5.5).

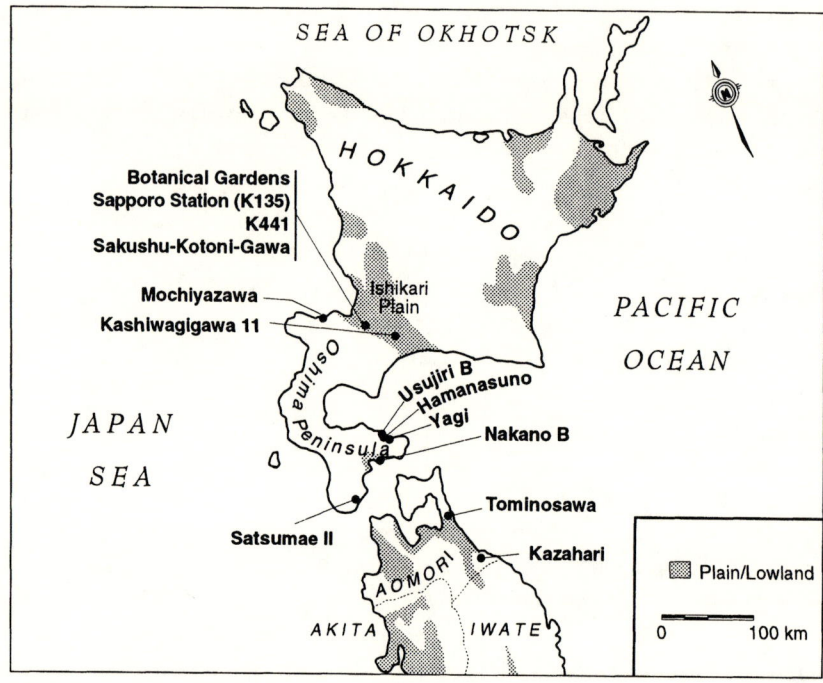

Figure 5.1. Map of northeastern Japan showing location of sites.

The interpretation of anthropogenesis at archaeological sites in northeastern Japan must, of course, hinge on a number of assumptions. Foremost among these is that plants indicative of ecological disruption attributable to human beings can be identified in the archaeological record. Plants that colonize human-disturbed habitats may also prefer habitats disturbed by natural influences such as tree-falls, riverine and coastal erosion, forest fires, and typhoons, which result in local plant communities of varying ages and stages of succession. In the following discussion, I limit interpretations to patterns shared by groups of sites. The patterns involve the relative abundance of annual weeds (indicative of regular, short-term disturbance) and perennial weeds (indicative of early successional stages that would be associated with forest edge or village edge communities). The frequency and intensity of disruptive activities represented by these plants encompass short-term disruptions, as well as actions that would maintain disequilibrium because of continual disturbance. Under these circumstances, ongoing human presence inhibited ecological succession. I do not assume that the weeds are *nec-*

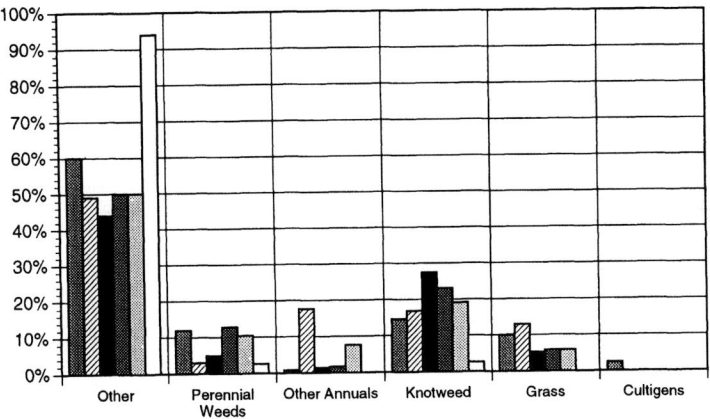

Kazahari (LJ)   Usujiri B   Tominosawa (MJ)   Hamanasuno   Yagi 2   Nakano B

Figure 5.2. Bar chart illustrating relative quantities of plant taxa for the Initial through Late Jomon. LJ = Late Jomon component; MJ = Middle Jomon component; IJ = Initial Jomon component; Yagi 2 = Yagi excluding "Other Grass." Bars are arranged chronologically from Initial Jomon (right) to Late Jomon (left).

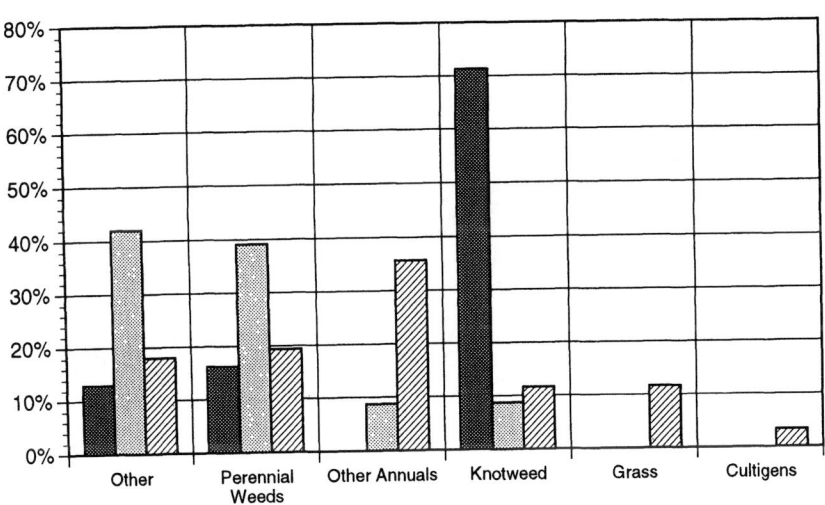

Sapporo Station (K135) Esan   Sapporo Station (K135) Kohoku   Mochiyazawa

Figure 5.3. Bar chart illustrating relative quantities of plant taxa (excluding nuts) for the Zoku-Jomon in Hokkaido.

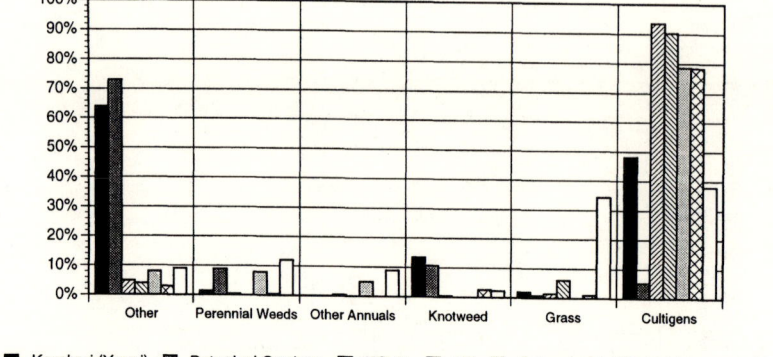

Figure 5.4. Bar chart illustrating relative quantities of plant taxa at six Ezo-Haji/Satsumon sites in Hokkaido and the Yayoi component of the Kazahari site. SK = Sakushu Kotoni River; KG-II = Kawshiwagi-Gawa-II.

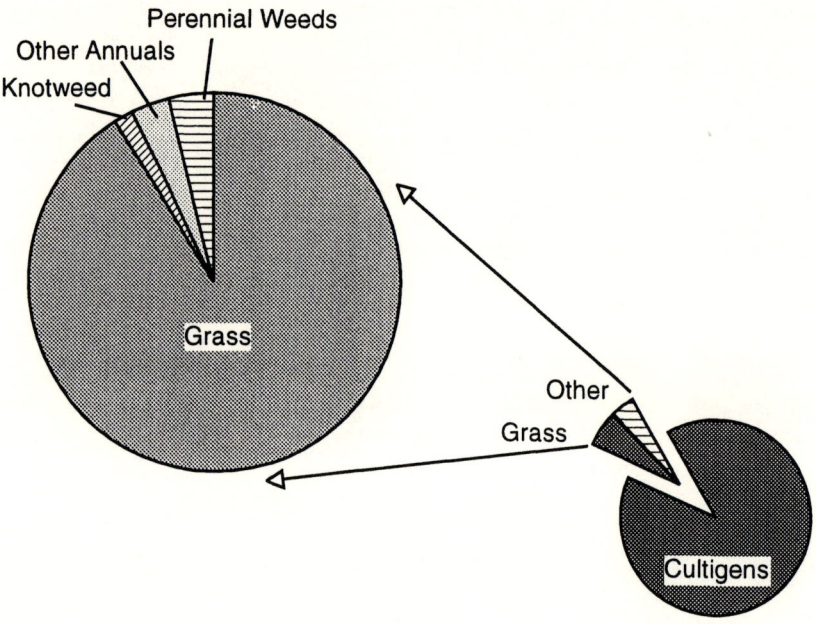

Figure 5.5. Pie charts illustrating the relative proportions of weed seeds from the Sakushu Kotoni River site, Hokkaido.

essarily a result of human influence. I argue, however, that this appears to be the case.

Initial Through Late Jomon (ca. 9500 to 3000 B.P.)

The Initial Jomon spectrum of plant remains is dominated by the "other" category of plants that comprises tree fruit including Amur corktree (*Phellodendron amurense*) and walnut (*Juglans ailanathifolia*) (figure 5.2) (Crawford 1983). Annual weeds and knotweed are barely present in the samples. Initial Jomon sites were communities with pit houses, but the hamlets/villages were usually much smaller and less densely occupied than those of subsequent Jomon communities. I am not inclined to interpret the nut procurement regime as anything other than harvesting from naturally occurring stands of walnut trees. Elsewhere I have argued that the evidence does not necessarily point to the nut tree management that Nishida (1980, 1981, 1983) proposes for central and southwestern Japan (Crawford 1992). The plant remains are consistent with a low degree of local ecological disturbance.

In contrast to the Initial Jomon pattern, the subsequent Early through Late Jomon pattern attests to a much greater degree of anthropogenesis. The plant remains assemblages from six sites are relatively consistent. Annual and perennial weedy plants range from about 36% to more than 50% of the seed assemblages (figure 5.2). Grasses, which have not been identified in the Initial Jomon samples, comprise a significant proportion of Early and subsequent Jomon assemblages. Most of these grass seeds are from a single genus, *Echinochloa* (barnyard grass). The other weed forming a high percentage of Jomon assemblages is the knotweed (*Polygonum*). Usually three or four species of *Polygonum* are present. Also present, particularly in the Middle and Late Jomon, is the closely related sheep sorrel (*Rumex* sp.). Then, by Late Jomon times, several cultigens, rice (*Oryza sativa*), foxtail millet (*Setaria italica* ssp. *italica*), broomcorn millet (*Panicum miliaceum*), and possibly buckwheat (*Fagopyrum esculentum*), were available in northeastern Japan in some quantity (Crawford 1992; D'Andrea 1992). One or two of these cultigens may have been present earlier (Crawford 1983; Crawford and Takamiya 1992), but they have not been recovered in the quantity that they have from the Late Jomon. The low of 36% weeds comes from Kazahari, the Late Jomon site with cultigens. A half dozen seeds of foxtail grass (*Setaria* sp.) in the Late Jomon assemblage at Kazahari, although not particularly striking, are consistent with the interpretation that cultivation was taking place. Foxtail grass is more common in flotation samples from the later agricultural phases in the northeast compared with its presence in the Initial through Middle Jomon phases there.

Small scale gardening may have taken place from the Early Jomon onward in northeastern Japan. Thus anthropogenesis in the form of clearing for gardens likely is responsible for some of the weed assemblage in the Late Jomon and may also account for some of these remains in sites as early as the Early Jomon. Besides gardening, clearing land for settlements, construction, firewood collection, and possibly burning may account for most of the anthropogenic influences upon plant communities that are reflected in the paleoethnobotanical record. Ecological disruption, then, developed as a consequence of the many activities associated with long-term village occupation and not necessarily or entirely in conjunction with gardening.

For now, we have not modeled how newly established communities procured their resources. In southwestern Hokkaido, for example, Jomon sites are quite large, often multicomponent, and number more than 90 within an area of less than 40 to 50 sq km of coastal terrace. Suitable resources, particularly those created by human influences, were likely not far away at any time (Bleed et al. 1989). Primary and secondary products of gardens likely provided some insurance for lean years as did secondary wild resource choices. Very few nut remains have been recovered from any of these sites.

Jomon people seem to have affected significantly the productivity of their terrestrial environment (Crawford 1983). Anthropogenic communities were sufficiently productive to encourage a de-emphasis on nut procurement. Nut remains from Jomon sites in southwestern Hokkaido decrease in density to the extent that, by the Middle Jomon, nuts are present but not regularly recovered in flotation samples. At Usujiri B they are barely present and at Tominosawa, too, acorn and walnut are present only in a few flotation samples (Crawford 1983; D'Andrea 1992). Acorn and walnut remains are commonly recovered from the Late Jomon and Yayoi components at Kazahari, however (D'Andrea 1992). Why nuts become more common late in the sequence when gardening becomes more intensive is a problem. Increasing human population density and increased clearance may be factors. Nut trees are more productive along forest edges and in openings, and nuts may have been valued more than they had been previously as a result of the investment of efforts in plant husbandry, which is a relatively risky undertaking because of the possibility of crop failure. Nuts, along with other gathered foods, would have cut these subsistence risks substantially. Plant husbandry could have been a factor in changing local ecology to suit human needs as early as the middle Holocene in northeastern Japan. In general, the Jomon cannot be assumed to have been passive extractors from local habitats.

Zoku-Jomon (ca. 2300 to 1600 B.P.)

The Zoku-Jomon, a phase between the Final Jomon and Ezo-Haji/ Satsumon of Hokkaido and a contemporary of the Yayoi, exhibits considerable variation in its seed assemblages. Three occupations representing two periods of this phase have been extensively sampled at the Sapporo Station (K135) and Mochiyazawa sites (Crawford 1992; Crawford and Takamiya 1992). The plant remains from these occupations are dominated by walnut and acorn (*Quercus* sp.). In general, perennial plants and other taxa are relatively common, more so than in the preceding Jomon periods (figure 5.2). They are probably more abundant than the chart indicates because one of the knotweeds, *Polygonum cuspidatum,* is a perennial. Annual plants such as grasses and knotweeds vary considerably in relative proportions. There is some question about the role food production played at each of these Zoku-Jomon sites (Crawford and Takamiya 1992; D'Andrea 1992). Cultigens are present at Mochiyazawa and Sapporo Station (K135). Barley (*Hordeum vulgare*) is represented by one specimen at Sapporo Station while twenty-one grains of rice, barley, foxtail millet, and broomcorn millet occur at Mochiyazawa. Mochiyazawa also has the highest percentages of weedy annuals and the only substantial representation of weedy grasses among the three Zoku-Jomon components reported here. None of these sites shows evidence of substantial house construction; thus although repeated occupation is evident, long-term continuous occupation of the sites may not have occurred. Mochiyazawa is one of many occupations in a 1 sq km area representing earlier Jomon phases as well as the more recent Ezo-Haji/Satsumon. Such repeated and long-term use of a limited area likely affected the local ecology. The pattern is difficult to interpret at the moment. One possibility is that the Mochiyazawa population was involved in food production, which would have been largely responsible for the weed communities there.

Food-Producing Phases (Postdating 1600 B.P.)

Four populations of substantial food producers are identified in the archaeological record of northeastern Japan: the Tohoku Yayoi, Ezo-Haji, Satsumon, and Heian. Heian is only tentatively identified at sites such as Uchiebisawa and Hachinohe. Their occupants may have been Ezo, although their material culture shared much with that of contemporary *wajin* (protohistoric Japanese) settlers in Tohoku. The Tohoku Yayoi are thought to have practiced rice production as did the Yayoi of southwestern Japan. However, this agricultural venture appears to have failed, although occupation of the area continued after adjustments were made to production methods. Instead of rice, Tohoku Yayoi populations came to rely on crops such as wheat, barley, and millet (Crawford and Takamiya 1992). The descendants of these people appear to be the Ezo, who are likely the ancestors of the Ainu. The Ezo relied on dryland food production (Crawford and Takamiya 1992; Crawford and Yoshizaki 1987).

Plant remains from the Tohoku Yayoi and Ezo-Haji are comprised of cultigens to a greater extent than those from any of the periods discussed so far (figures 5.4 and 5.5). Cultigen seeds usually represent more than 40% of these assemblages, including the Yayoi component at Kazahari, with the exception of the Botanical Gardens site where they comprise only 5% percent of the collection. The pattern of annual and perennial weeds is similar to that of the Early through Late Jomon in that grasses and knotweeds are common, but their relative abundance is lower because of the significant quantities of cultigen seeds. Goosefoot (*Chenopodium* sp.) and sheep sorrel (*Rumex* sp.) are common annuals, as they are in some Jomon assemblages. Sumac (*Rhus* sp.) and fleshy fruit-bearing bushes are also well represented. Differences include a greater diversity of archaeological grasses compared with those in Jomon collections. Barnyard grass (*Echinochloa crus-galli*) is present, but the most common of the weedy grasses is usually a foxtail grass (*Setaria* sp.). Very few of the knotweeds are *Polygonum sachalinense,* which is common in the Jomon samples; instead, they are usually *P. densiflorum*-type or *P. lapathifolium.*

The Grasses

Details of some of the archaeologically recovered grasses suggest further insights into anthropogenesis in prehistoric northeastern Japan.

Grasses are an extremely diverse plant family (*Gramineae* or *Poaceae*). Assemblages from the sites in question, however, are marked by very few grass species. The predominant representatives among the archaeological grasses in northeastern Japan belong to the tribe *Paniceae*, a group known for its economic value to people. The common members of the *Paniceae* in the Japanese samples are barnyard grass and barnyard millet (*Echinochloa crus-galli* and *E. utilis,* respectively) and foxtail grass and foxtail millet (*Setaria italica* ssp. *viridis/glauca* and *S. italica* ssp. *italica,* respectively).

Barnyard Grass (Including Rice Paddy Weed) and Barnyard Millet

Barnyard grass and barnyard millet are two species of *Echinochloa* that appear to have a long history of association with people in Japan. The first species is a weed (barnyard grass, *E. crus-galli*) with two distinct forms, dryland (*E. crus-galli* var. *crus-galli*) and rice paddy–associated varieties (*E. crus-galli* var. *oryzicola*). The second species is a cultigen (barnyard millet, *E. utilis*). The three taxa belong to the same genome, one that does not include the South Asian cultigen (*E. frumentacea*) (Yabuno 1966). Researchers grow about 120 cultivars of barnyard millet at the National Northeast Agricultural Experiment Station in Tohoku, Japan (Yabuno 1987:486). *Echinochloa crus-galli* is a successful weed for a number of reasons, including its production of large numbers of seeds per plant, seed dormancy, and its capacity to grow on a wide variety of soil types and textures (Maun and Barrett 1986).

Jomon, Yayoi, Ezo, and Ainu plant remains assemblages clearly evidence several types of human-*Echinochloa* associations that represent distinct activities. The identification of these associations depends on our ability to distinguish the three taxa in archaeological plant remains assemblages. The cultigen caryopses (grains) are significantly larger than those of weedy *Echinochloa*. The dorsal surface of the cultigen has a distinct hump at the point where the steeply inclined embryo terminates near the middle of the grain, whereas weedy *E. crus-galli* var. *crus-galli* grains have a flat dorsal surface and the seeds are markedly smaller than the cultigen grains. The grains of *E. crus-galli* var. *oryzicola* are slightly larger and two to three times more massive than the dryland weed grains (Maun and Barrett 1986). The embryo of *E. crus-galli* var. *oryzicola* caryopses extends the whole length of the seed rather than terminating at half to three-quarters the length of the seed as it does in *E. crus-galli* var. *crus-galli*. Scanning electron microscope examination of the epidermal surface of the grains and the fruit has so far provided no other characters that distinguish the three forms.

Barnyard millet (*E. utilis*) has been identified in substantial quantities from a 400–300 B.P. Ainu house in Nibutani, western Hokkaido. A few grains of barnyard millet have also been tentatively identified from one or two Ezo sites dating from 1200 to 800 B.P. These Ezo sites generally have quantities of broomcorn millet grains, which are similar to barnyard millet grains in size and general shape. One site with thousands of grains of broomcorn millet (Sakushu Kotoni River) may have many barnyard millet seeds as well, but this finding is difficult to confirm. At the Megumi site in Yonago city, southwestern Japan, barnyard millet seeds have been identified in both Yayoi and Final Jomon components (Kasahara et al. 1986:121). In any case, barnyard millet grains are being identified in contexts that, on the basis of other archaeological evidence, are known to be agricultural.

Echinochloa crus-galli var. *oryzicola* is a rice-associated weed that developed as a rice-plant mimic (Barrett 1983, 1987). The rice mimic is associated only with paddy-field agriculture. Until recently, no examples of seeds of this plant had been located in the archaeological record. In the past few years the situation has changed. The Okawa site, located on the Sea of Japan coast of Hokkaido in Yoichi, has yielded a number of clear examples of *E. crus-galli* var. *oryzicola,* as well as a few grains of cultigen barnyard millet (*E. utilis*). In addition, rice grains have been recovered from the same site in several localities. A radiocarbon date on the rice from Okawa is 780 ± 200 B.P. (TO-1999), and another date on broomcorn millet grains from the site is 1160 ± 80 B.P. (TO-1998). Calibrated ages for these materials are A.D. 1280 and A.D. 890, with calibrated age ranges at the 68.3% confidence interval falling between A.D. 1028 and 1400 and A.D. 782 and 984, respectively (using CALIB 3.0, Stuiver and Reimer 1993). The houses here are not characteristically Ezo in style, so Okawa may be a *wajin* occupation. A small collection of Ezo pottery has been retrieved from Okawa, however. Seeds of *E. crus-galli* var. *oryzicola* have been recovered from only one other site in the study area. Forty-one grains come from pits at Mochiyazawa, a Zoku-Jomon site. Thirteen of these grains are measurable, with an average size of 1.9 (1.8–2.2) mm long, 1.3 (0.8–1.7) mm wide, and 1.1 (0.7–1.7) mm thick (D'Andrea 1992).

Rice has been recovered from three Ezo period sites, including

Okawa, as well as from the preceding Zoku-Jomon Mochiyazawa site in Hokkaido. The rice-paddy weed is also found at two of these sites. At the moment, distinguishing whether rice was grown at, or imported to, these sites is difficult to determine. No plant parts other than the grains have been found. If the rice was imported, the rice-mimic barnyard grass seeds likely came with it. Nonetheless, rice may well have been grown at Okawa and Mochiyazawa; the presence of the paddy weed is circumstantial evidence that the rice was locally produced.

Barnyard grass seeds (*E. crus-galli* var. *crus-galli*) are present at most sites with grass remains, including Jomon and later sites. Small numbers are present at Early Jomon sites, while larger quantities are found at Middle and Later Jomon sites, and the fully agricultural components of the period 1600–1000 B.P. in Hokkaido have abundant barnyard grass remains. The most detailed study of archaeological barnyard grass is the analysis of specimens from the Hamanasuno and Usujiri B sites on the Kameda Peninsula in southwestern Hokkaido (Crawford 1983). Three hundred twenty-one grains from the Early and Middle Jomon periods have been identified in flotation samples from the two sites. Only thirty-three samples from Hamanasuno are from floor or pit/hearth contexts. Except for one specimen, the barnyard grass seeds are all from pit/hearth flotation samples. The majority (139 seeds) came from a post hole in House 72 at Hamanasuno. The Tominosawa site samples are only from house floors and features, so the occurrence of the grass in other contexts cannot be ascertained. At Usujiri B most of the shallower levels at the site had been destroyed before the excavation began, so any stratigraphic variation in the occurrence of the grass seeds can also not be explored. Barnyard grass seeds occur in floor, pit, burial, and deep house fill (Level X3: the stratum normally immediately overlying the few centimeters of floor fill) contexts. In addition, a carbonized mass adhering to a Middle Jomon pot sherd contained *Echinochloa* seeds. In general, barnyard grass seeds occur only in low densities except at one or two locations at these sites. Their regular occurrence at the sites and their associations primarily with pits, hearths, floors, and, in one case, residue in a pot support the contention that the seeds were collected and used, likely as food, by the site inhabitants.

The *Echinochloa* seeds are not particularly abundant at any of the sites (in fact, densities for seeds of all taxa are low). However, despite the apparent nonintensive use of this grass, indications are that a phenotypic change had occurred by 4000 B.P. (Crawford 1983). The mean size of the grains increased about 20% by the end of the Middle Jomon (ca. 4000 B.P.). Furthermore, nearly one-quarter of the *Echinochloa* seeds from the Middle Jomon Usujiri B site are larger than any of the grains found at Hamanasuno. Only one specimen from Hamanasuno is longer than 2 mm, but it is relatively narrow. Finally, the frequency distributions of the grain measurements indicate at least two modes and therefore two populations of *Echinochloa* distinguishable by grain size (Crawford 1983). I will be examining the details of the archaeological *Echinochloa* variation elsewhere (Crawford, forthcoming).

A specimen not included in the statistical analysis of the Jomon *Echinochloa* grains came from the carbonized mass attached to a rimsherd recovered from a house floor adjacent to a hearth. The specimen is a nearly complete grain with palea and lemma attached. The grain size is outside the range of weedy *Echinochloa* and within the range of *Echinochloa utilis,* the cultigen barnyard millet. The grain also has the distinctive hump in the central region of the dorsal surface found in the cultigen but not in the weed.

The large grains from Usujiri B are within the size range of both the rice paddy–mimic and cultigen species of *Echinochloa*. The sixty varieties of barnyard millet have variable grain size. The length varies from 1.6 to 2.8 mm and the width varies from 1.7 to 2.3 mm. Experiments show that *Echinochloa* caryopses shrink by as much as 10% during carbonization (Crawford 1983:31). Thus more than 25% of the Usujiri B barnyard grass grains are within the range of the cultigen species. The embryo shape of all the grains and the dorsal surface contour on the well-preserved specimen are not at all like the respective features of the rice paddy–mimic form of barnyard grass. The grains are very similar to the cultigen species of *Echinochloa*.

Two cultigen populations in North America, sunflower (*Helianthus annuus*) and sumpweed (*Iva annua*), exhibit an increase in seed/achene size through time (Asch and Asch 1978; Black 1963; Yarnell 1972, 1978). Such an increase in seed size often accompanies domestication of a species (Schwanitz 1967:14–24). The characteristics of the Japanese *Echinochloa* grains indicate that phenotypic variation attributable to selection processes leading to domestication were under way in Hokkaido.

Other Millets: Foxtail and Broomcorn

About a half-dozen grains of a spherical grass grain were recovered from a flotation sample at Usujiri B. They are morphologically identical to domesticated *Setaria italica* ssp. *italica,* foxtail millet. No other foxtail millet grains have been recovered from the site, so their meaning is difficult to assess. They have not been recovered from the only other extensively sampled and analyzed Middle Jomon site in northeastern Japan, Tominosawa (D'Andrea 1992). Two possibilities may be entertained. First, the grains may be intrusive from a later period. Some examples of Final Jomon pottery occur in shallow house fill levels. Ainu material, including a burial, has been found at Usujiri B. However, none of these is associated with the sample in question. No other cultigen remains characteristic of Ezo or Ainu subsistence occur at Usujiri B. Second, the specimens may be Jomon. If so, they are further corroboration of the presence of cultigens in northeastern Japan during the Jomon.

The only other known examples of millets from northeastern Jomon sites come from the Kazahari site in the Aomori prefecture. D'Andrea (1992) has identified both foxtail millet (*Setaria italica* ssp. *italica*) and broomcorn millet (*Panicum miliaceum*) from the Late Jomon component of the site. Rice comes from the same component and has been dated using the accelerator mass spectrometry technique to the early third millennium B.P. (D'Andrea et al. 1995). The evidence is building for gardens being present during the Jomon of northeastern Japan.

There is no evidence that either foxtail millet or broomcorn millet was domesticated locally. However, a few examples of foxtail grass, probably green foxtail (*Setaria italica* ssp. *viridis*) and panic grass (*Panicum* sp.) are found in Jomon flotation samples, although they are relatively rare. Two specimens of probable panic grass have been recovered from Hamanasuno site floor and pit samples (Crawford 1983:35). Only one probable foxtail grass has been found in samples from the Kameda Peninsula Jomon; this was found in a Middle Jomon house (House 62) at Hamanasuno. D'Andrea (1992) reports a similar single occurrence of foxtail grass from the floor of House 87 at Tominosawa. D'Andrea also found one unidentifiable *Paniceae* seed from the same house floor. In the Late Jomon contexts at the Kazahari site in the Aomori prefecture are the six foxtail grass seeds mentioned earlier. Wild foxtail grass (likely green foxtail) and probable panic grass are well represented in assemblages from later sites such as Sakushu Kotoni River, which have ample evidence of crops. Quite likely in this last case these grasses are field weeds, and the same may be true for the Late Jomon Kazahari occupation. Green foxtail, in particular, is a very common grain-field weed

(Douglas 1985). Two weedy taxa of panicoid grasses, *Echinochloa* and *Setaria,* are common in Ezo period sites. They are likely garden weeds, and the seeds are found in very low densities compared with the occurrence of cultigens from the same sites.

Other Grasses

At least twenty other grass taxa occur in the Kameda Peninsula Jomon samples. Not all have been identified, however. Rye grass or wheat grass seeds (*Elymus* sp. or *Agropyron* sp.) occur in samples from the Yagi site in southwestern Hokkaido. Ninety-four seeds have been found, and most come from one sample, an occupation floor within the fill of House 3. The grains are quite large (3.8–4.4 mm long). Also in the Yagi samples are more than a thousand yet unidentified small (1.0–1.5 mm) grass seeds. Most of these are from House 3, and their occurrence has a strong positive correlation with the occurrence of the *Elymus* or *Agropyron* seeds. A different type of small grass seed occurs in a hearth of House 60 and in House 30A at Hamanasuno (Crawford 1983:35).

Discussion

Comparison of carbonized plant remains assemblages spanning some seven millennia in northeastern Japan provides evidence of four weed associations: one for the Initial Jomon, another spanning the Early through Late Jomon, a Zoku-Jomon pattern, and a clear agricultural pattern during the Yayoi and Ezo-Haji/Satsumon. Localized processes of anthropogenesis are distinguishable in the archaeological sequence much earlier than regional pollen studies suggest. Early through Late Jomon exhibits a relatively consistent pattern of annual and perennial weeds while the subsequent periods, the Zoku-Jomon and Yayoi-Ezo, have weed assemblages consistent with wet rice production and dryland plant husbandry, as well as seeds of domesticated species/plant staples. Evidence points to plant domestication and cultivation earlier than the Late Jomon and certainly by the end of the Late Jomon in northeastern Japan.

The weedy grasses from the Jomon are consistently represented by *Echinochloa.* With the advent of intensive food production, *Echinochloa* is still present, but *Setaria italica* ssp. *viridis* is found in significant quantities. When rice is found in abundance at sites, the rice mimic, *E. crusgalli* var. *oryzicola,* is also present. The one exception is the Late Jomon occupation at the Kazahari site where rice is associated with foxtail millet and broomcorn millet but with no evidence of paddy-field production methods. Weedy grasses are low in diversity at these sites, as they are in the Jomon. Although these grasses are higher in density than in the preceding Jomon samples, their proportion in relation to other seeds is low because of the quantities of cultigen grains. *Echinochloa* and *Setaria italica* ssp. *viridis* are likely weeds of tilled fields and represent contamination of grain harvests (barley, wheat, millet) rather than purposeful gathering.

By the end of the Middle Jomon, *Echinochloa crus-galli* grains appear to undergo a change to a form that is indistinguishable from the cultigen species, *E. utilis.* So far, evidence for this form comes from only one site, but flotation sampling at Middle Jomon sites is still rare. Furthermore, grasses were apparently part of a harvesting regime during the Jomon. It is within this context that domestication may have been occurring.

At the outset I argued that anthropogenesis is an issue that must be examined if we are to have any understanding of the longevity of the Jomon and the changes that led to the beginning of plant husbandry as a significant aspect of aboriginal life in northeastern Japan. Systematic research during the past twenty years on plant assemblages in northeastern Japan is helping us come to an understanding of these issues. Pollen analysis is but one component of the research focusing on human-plant interactions in Japan. Assemblages of plant remains recovered by flotation provide a much more localized indication of the ecological contexts that were useful to people. With continued research we will be able to develop a much clearer picture of the reciprocal relationships that obtained between human beings and plants in Japan and the roles that people played in the biodiversity of this Asian archipelago.

PART 2

Plant Resources, Human Communities, and Anthropogenic Landscapes

CHAPTER 6

Between Farmstead and Center
The Natural and Social Landscape of Moundville

C. MARGARET SCARRY AND
VINCAS P. STEPONAITIS

Between A.D. 900 and 1650 the Black Warrior Valley of Alabama was the setting for dramatic cultural changes. This period encompassed the emergence, florescence, and dissolution of the Moundville polity. At the beginning of the period, the valley was inhabited by people who lived in egalitarian communities and relied on foraging and small-scale gardening for their food. Over the next century or two, the people of the Black Warrior Valley reorganized their social, political, and economic relations. The changes they made ultimately resulted in the establishment of a complex chiefdom supported by an economy based on corn agriculture. This polity was controlled by a chief, who lived at the site we know today as Moundville (figure 6.1).

We and our colleagues, both past and present, have devoted considerable effort to investigating the emergence and organization of the Moundville chiefdom and to defining its place in the Mississippian world of the late prehistoric Southeast (see for example Bozeman 1982; Jones and DeJarnette n.d.; Knight 1989; McKenzie 1964, 1966; Moore 1905, 1907; Peebles 1974, 1983, 1987a, 1987b; Powell 1988; Scarry 1986; Steponaitis 1983, 1991, 1992; Welch 1990, 1991). Here, however, we are going to examine the Moundville polity from a different perspective. Our focus will be the dynamic period between A.D. 900 and 1250, when the Moundville polity was taking shape. First we will outline the development of the polity and describe changes in the production and procurement of plant foods that accompanied the changes in social and political relations. Then we will discuss economic relations between people living at Moundville and people living on farmsteads. Finally,

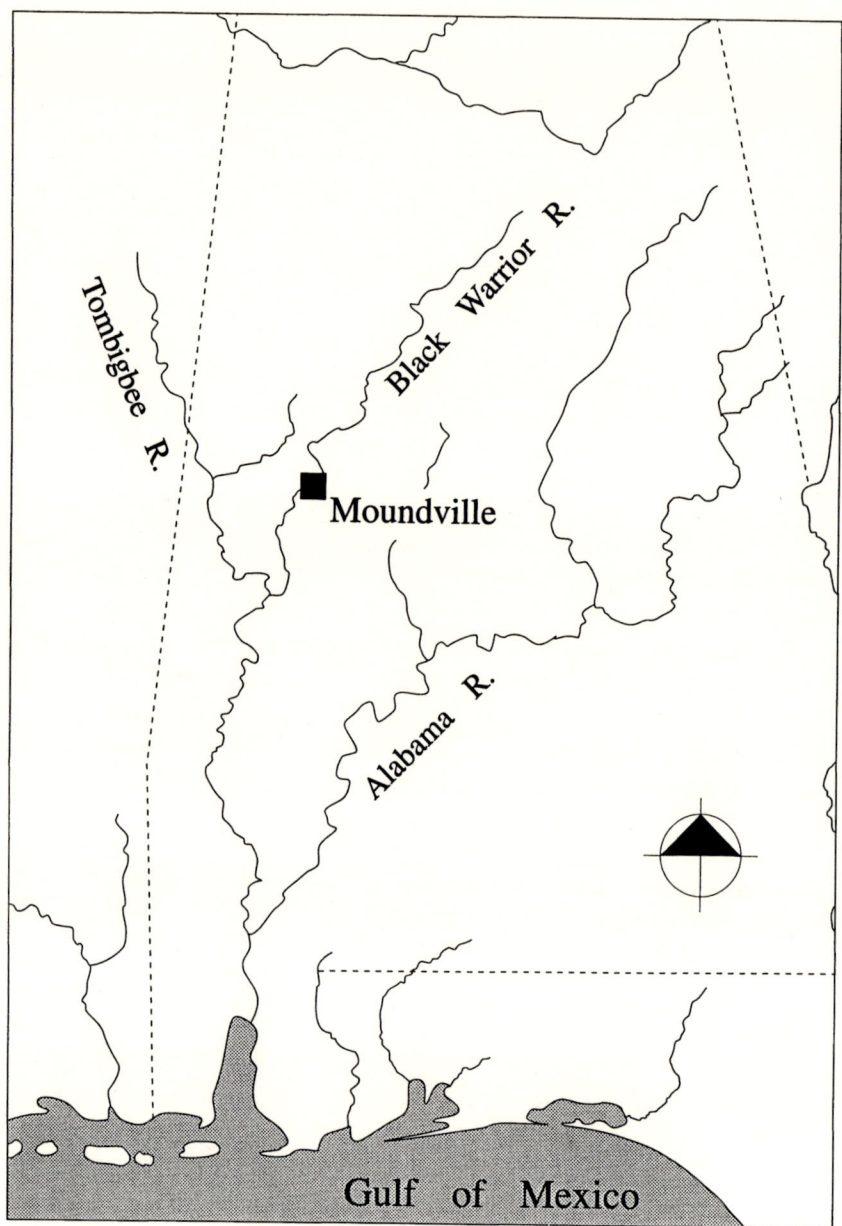

Figure 6.1. Location of the Moundville polity.

we will offer some thoughts about how the changes in the cultural landscape affected the natural landscape.

Social and Political Development of the Moundville Polity

The late prehistoric occupation of the Black Warrior Valley can be divided into five chronological phases, all but the last of which can be further subdivided into early and late subphases based on differences in ceramic styles (Curren 1984; Jenkins and Nielsen 1974; O'Hear 1975; Steponaitis 1983, 1992). The phases provide a framework for tracking trends in settlement, political, and economic organization.

During the West Jefferson phase, A.D. 900–1050, people lived in small villages that were scattered over favorable locations on the valley's floodplain. Evidence for political differentiation among West Jefferson communities is nil (Welch 1990). The sites do not have mounds and are not known to contain elaborate burials. The overall situation seems to have been one of autonomous villages and a relatively egalitarian society.

With the beginning of the Moundville I phase, at about A.D. 1050, the social landscape changed dramatically. Single pyramidal mounds were constructed at several of the former villages. At roughly the same time, most people left the nucleated villages and began living in dispersed farmsteads, which consisted of one or two households surrounded by their fields. Presumably each mound site served as the political, economic, and ritual focus for the people living on farmsteads nearby (Bozeman 1982; Steponaitis 1983).

So far as we know, the early local centers may have been roughly equivalent in terms of the number and size of mounds. The center at Moundville, however, differed significantly from the others. It had an unusually high number of people living in its immediate vicinity. Recent analyses of sherd collections indicate that approximately three-quarters of the chronologically diagnostic sherds from midden deposits (as opposed to mortuary or mound contexts) at Moundville date to the Moundville I phase (Steponaitis 1991, 1992; Welch 1989). These data suggest that Moundville's resident population was greatest in this early phase. This picture is reinforced by the results of salvage excavations conducted on the northwest edge of the site by the Alabama Museum of Natural History. In an area of approximately 1800 sq m, remains of at least thirteen structures were found (Scarry 1995). Ceramics and a series of radiocarbon determinations indicate that the structures are predominantly, if not exclusively, Moundville I in date. The full extent and distribution of Moundville I middens at Moundville are unclear. Discrete patches of midden seem to be scattered across the terrace on which the site is located. This suggests to us that an unusually high density of farmsteads once dotted the area. Based on surveys of the valley that have been done to date, no other district had as high a concentration of Moundville I settlements. We would stress that Moundville had no obvious advantage over the other centers in the abundance or fertility of nearby soils. Thus the causes of this centripetal tendency must have been social and political, rather than purely environmental (Steponaitis 1991).

Whatever the case, during the Moundville I phase at ca. A.D. 1150, a second major transformation occurred, as Moundville gained clear political ascendancy over the entire region. The most obvious expression of this change was the construction of an enormous civic-ceremonial precinct covering some 100 ha. This precinct consisted of some twenty pyramidal mounds arranged around a large rectangular plaza (figure 6.2). The complex was enclosed on all but the bluff edge by a massive, bastioned palisade. (Note that the recent excavations on the riverbank at Moundville produced evidence that the palisade dates to the Moundville I phase [Scarry 1995].) It is clear from the site's symmetrical structure that the layout was deliberate and that the position and size of the mounds had social and religious meaning (Knight 1989; Steponaitis 1991).

Some 150 years after the mound-and-plaza complex was built, Moundville's resident population greatly declined, as indicated by the relative paucity of middens dating to the Moundville II and III phases (Knight 1989; Steponaitis 1991; Welch 1989). The smaller contingent of people who continued living at Moundville after A.D. 1300 probably comprised the pinnacle of the region's social, political, and religious elite.

The Moundville II and III phases, A.D. 1250–1550, encompass a time when the chiefdom was entrenched and Moundville served as a center of politics and ritual for the entire region. Analyses of burials from Moundville indicate that social differentiation was pronounced. There is also substantial evidence that Moundville's elite had ties with other

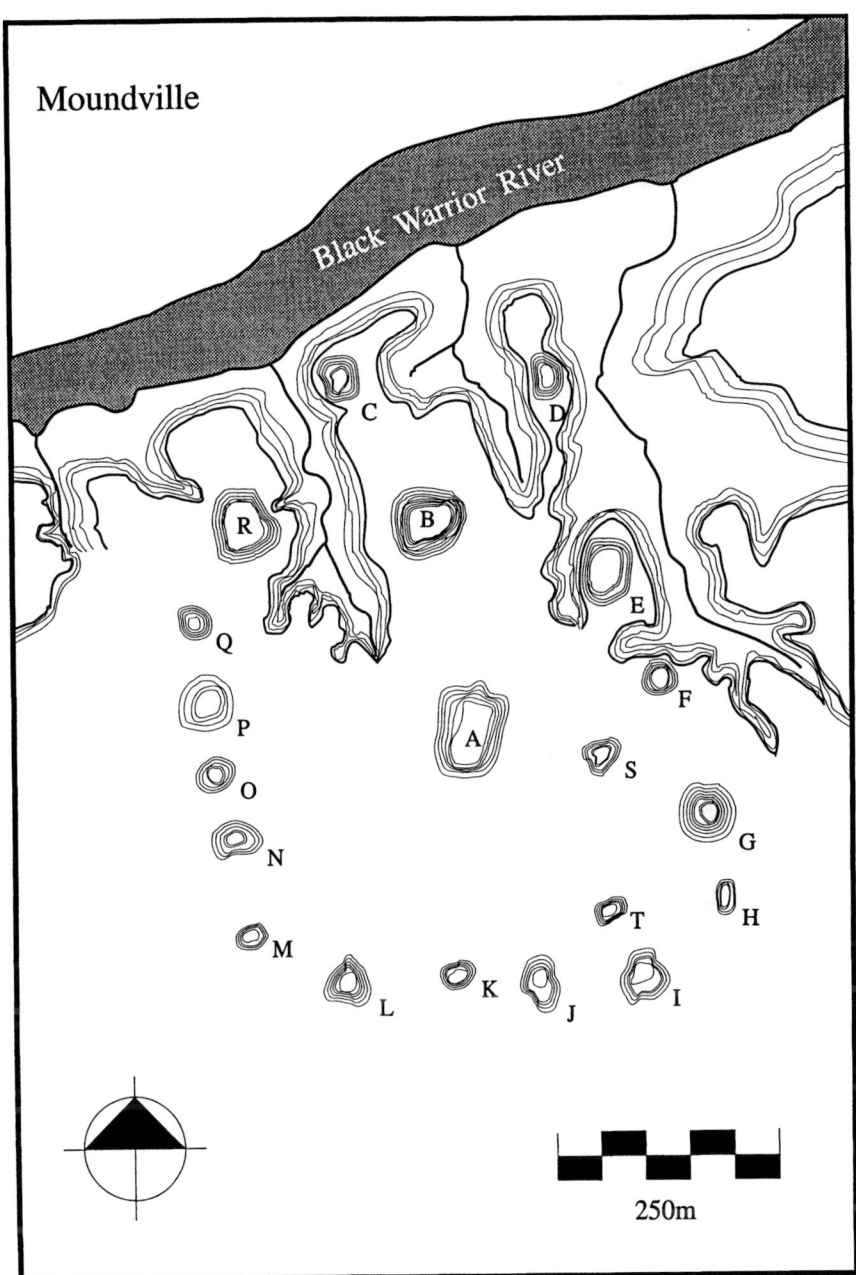

Figure 6.2. The mound-and-plaza complex at Moundville.

polities through long-distance exchanges of prestige goods (Peebles 1974, 1987a, 1987b; Steponaitis 1991; Welch 1991).

Although Moundville's elite clearly dominated the social and political scene in the Black Warrior Valley, other local centers (each with only a single mound) continued to be used. Presumably these centers were under the control of local chiefs, who though subordinate to the paramount at Moundville had jurisdiction over their immediate districts. As before, most of the region's population lived in dispersed farmsteads, where the machinations of the elite may have had little impact on their daily lives.

Sometime after A.D. 1500, the chiefly superstructure began to unravel (Peebles 1986; Steponaitis 1991). The dissolution of the polity was rapid. By the beginning of the Moundville IV phase, A.D. 1550, Moundville and the single-mound centers were no longer in use. People were once again living in nucleated villages, and all evidence of ranking disappeared from burials.

Agricultural Economy of the Moundville Polity

Having sketched the development of the Moundville polity, let us now examine plant production and procurement in the West Jefferson and Moundville I phases. The patterns we describe are derived from analyses of food plant remains that were recovered by flotation. The West Jefferson data came from twenty refuse pits from seven sites dating to early and late West Jefferson times (Scarry 1986:132–38). Moundville itself is represented by remains from thirty-two refuse pits that date to the late Moundville I phase and come from elite residential deposits (Scarry 1986:138–74). Finally, we have data from sixteen refuse pits from two Moundville-era farmsteads (Scarry 1993). The ceramics from one farmstead suggest that it dates to the Moundville I phase (Michals 1990). The other farmstead lacks diagnostic ceramics but has indications that it also was occupied in Moundville I times (Ensor 1993).

The food plants used by the people of the Black Warrior Valley were essentially the same in all contexts we examined (see Scarry 1986, 1993 for detailed analyses of the data summarized below). Corn (*Zea mays* ssp. *mays*), hickory nuts (*Carya* sp.), and acorns (*Quercus* sp.) were the dominant resources. Remains of these plants occurred in virtually every sample. Native crops and wild fruits also were used. Seeds from these plants, however, were neither as abundant nor as frequent in the samples as corn and nut remains.

People relied on the same resources in West Jefferson and Moundville I times and at the center and on farmsteads. Nevertheless, there were significant differences in the abundances of the three major resources in different settings. In the following section, we will first describe temporal patterns of food plant distribution seen in the West Jefferson data and the Moundville I data from Moundville. Then we will describe spatial patterns of plant use by examining data from West Jefferson contexts and from Moundville I contexts at Moundville and the two farmsteads.

The patterns of plant abundance are illustrated here with boxplots (Tukey 1977; Velleman and Hoaglin 1981:65–81; Wilkinson 1990:165–71), which are used to display and compare the frequency distributions of counts for different taxa. Each boxplot consists of several elements whose positions along a scale correspond to key points in the distribution: (a) the vertical line within the box indicates the median; (b) the vertical lines at either end of the box mark the "hinges," which approximate the 25th and 75th percentiles; (c) the lines extending outward from either end of the box, commonly called "whiskers," encompass the tails of the distribution; and (d) the free-standing asterisks and circles beyond the whiskers indicate anomalous "outliers" and "far outliers," respectively. In addition, each boxplot is marked by a "notch," where the box is constricted like an hourglass. The notch defines a simultaneous 95% confidence interval around the median; if the notches of two boxplots on the same graph do not overlap, then the sample medians are significantly different at the 0.05 level. Sometimes, when the notch extends beyond the hinge, the box acquires a curious appearance, as though it were bent back on itself; this does not change the way in which the graph is interpreted.

We should also note that, for present purposes, the distributions being compared are not of raw counts, but rather of counts that are modified in two ways. First the raw counts are divided by the total weight of plant remains from the same provenience, thereby producing "standardized" ratios that correct for differences in gross sample size. Then the standardized ratios are reexpressed computing their natural logarithms (adding one before taking the logarithm permits inclusion of samples that have counts of zero). This mathematical transformation has a number of advantages, not the least of which is that it "normalizes"

C. M. Scarry & V. P. Steponaitis

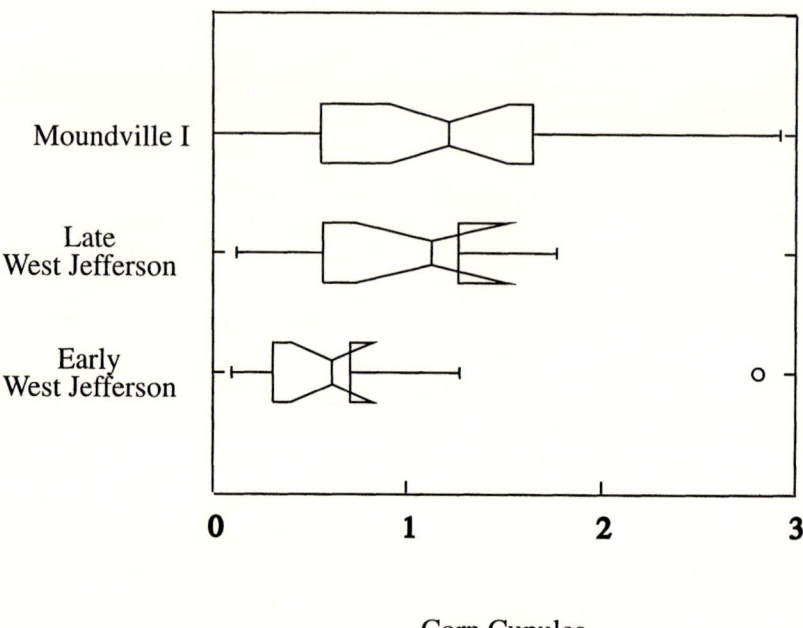

Corn Cupules

Figure 6.3. Boxplots comparing relative abundances of corn cupules in West Jefferson and Moundville I contexts. Values plotted are standardized counts reexpressed as natural logarithms (see text).

skewed distributions and thereby facilitates the visual (and statistical) recognition of patterns in the data (Doran and Hodson 1975:19, 49–50, 127; Velleman and Hoaglin 1981:48–55; Cleveland 1985:104–14). Thus our standardized, log-transformed counts are calculated as $\ln[(c/w) + 1]$, where, for each provenience, c is the count of the taxon in question and w is the weight in grams of all plant remains. These are the values plotted in all of the graphs that follow.

When we examine the relative abundance of corn cupules in different temporal contexts, we see that the intensity of corn production changed over time (figure 6.3). There was a significant increase in the production of corn from early to late West Jefferson times. On the other hand, levels of corn production appear comparable in late West Jefferson villages and from late Moundville I deposits at the paramount center.

While corn production increased, procurement of nuts decreased. The timing of this change, however, was different from that of corn

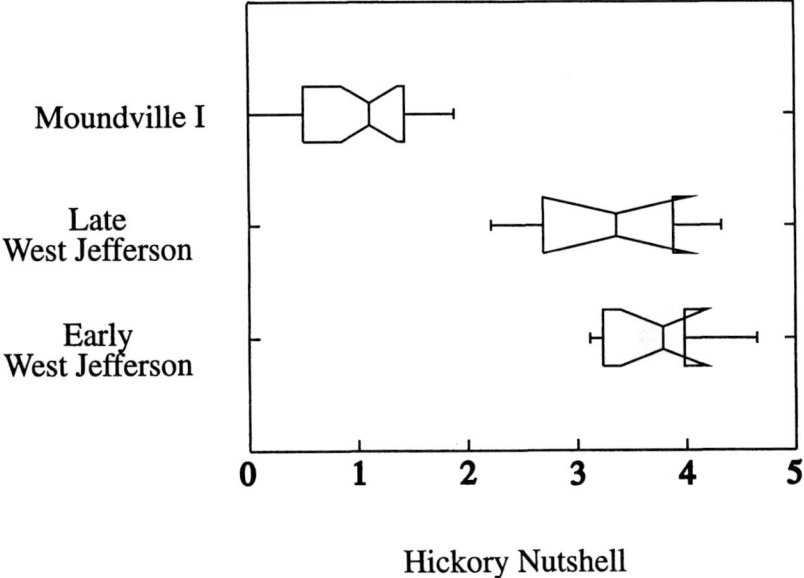

Hickory Nutshell

Figure 6.4. Boxplots comparing relative abundances of hickory nutshell in West Jefferson and Moundville I contexts. Values plotted are standardized counts reexpressed as natural logarithms.

production. Within the West Jefferson phase hickory-nut use was relatively stable (figure 6.4). Between late West Jefferson and Moundville I there was a dramatic drop in hickory-nut use. The same pattern can be seen in acorn use, although the decrease in acorn use was not as extreme as that of hickory nut (figure 6.5).

The plant data from West Jefferson and Moundville I contexts indicate that, between A.D. 900 and 1250, people in the Black Warrior Valley altered their subsistence strategies. The early West Jefferson assemblage suggests an economy based on foraging combined with small-scale gardening; people grew some corn and small quantities of native seed crops, but their primary plant foods were nuts. The increase in corn production during the West Jefferson phase and the decrease in nut procurement between the West Jefferson and Moundville I phases suggest an economy in transition. By the time Moundville had achieved political dominance in the valley, an economy focused on corn agriculture seems to have been firmly established (Scarry 1986).

At this point we need to make a caveat. We are monitoring changes

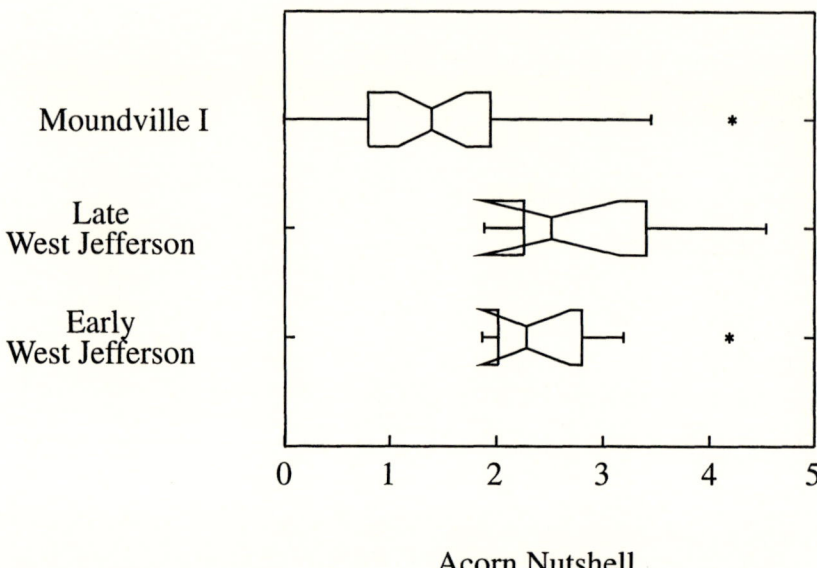

Acorn Nutshell

Figure 6.5. Boxplots comparing relative abundances of acorn nutshell in West Jefferson and Moundville I contexts. Values plotted are standardized counts reexpressed as natural logarithms.

based on differential quantities of corn cupules and nutshell fragments. These are inedible by-products. They may have been discarded when food was prepared for consumption. If so, presumably they would be deposited near where the food was cooked and served. But cobs and nutshells may also have been discarded when food was processed for storage or transport. Edible portions—shelled corn, hickory oil, or acorn meal—could have been transported, stored, and consumed elsewhere. In other words, we are measuring levels of production and processing, not consumption.

This distinction between processing and consumption becomes important when we add the farmsteads to the picture. The boxplots in figure 6.6 compare quantities of corn cupules from West Jefferson villages, the Moundville-era farmsteads, and the late Moundville I deposits at Moundville. The farmsteads have significantly more corn cupules than the West Jefferson villages *and* significantly more cupules than the deposits at Moundville.

The quantities of nutshells recovered from the various settings also

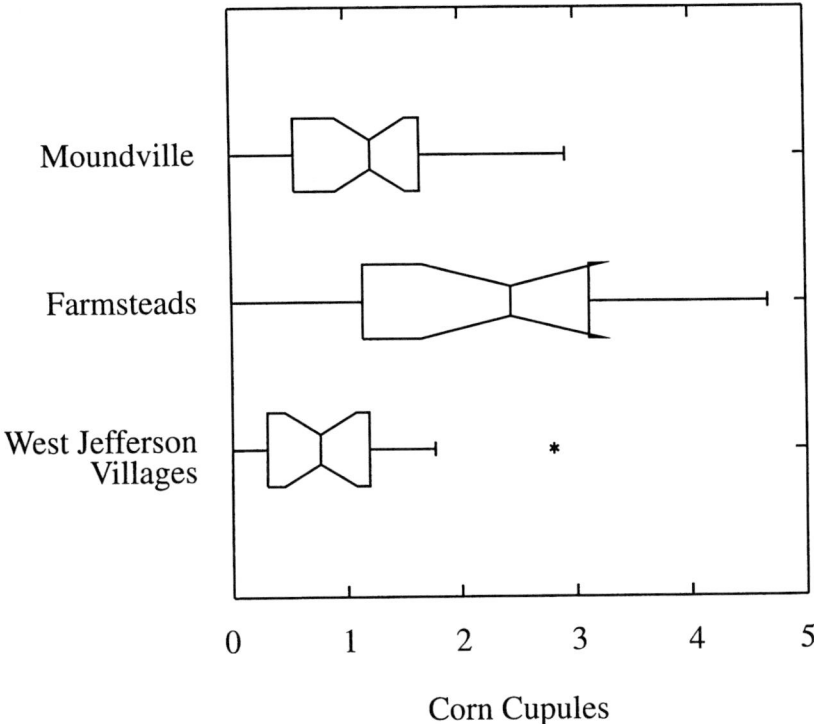

Figure 6.6. Boxplots comparing relative abundances of corn cupules at Moundville, the farmsteads, and West Jefferson villages. Values plotted are standardized counts reexpressed as natural logarithms.

show interesting patterns. The farmsteads yielded quantities of hickory nutshell that were roughly comparable to those from West Jefferson villages and significantly greater than those from Moundville (figure 6.7). The pattern is similar for acorn nutshell: West Jefferson villages and the farmsteads yielded similar quantities of acorn nutshell, and at both there were significantly more acorn nutshells than at the paramount center (figure 6.8).

    In sum, people at the farmsteads were producing or at least processing more corn than their predecessors in the West Jefferson villages and more corn than their elite contemporaries at Moundville. The residents of the farmsteads were also procuring and processing nuts at levels comparable to those of their predecessors and greater than those of their elite contemporaries.

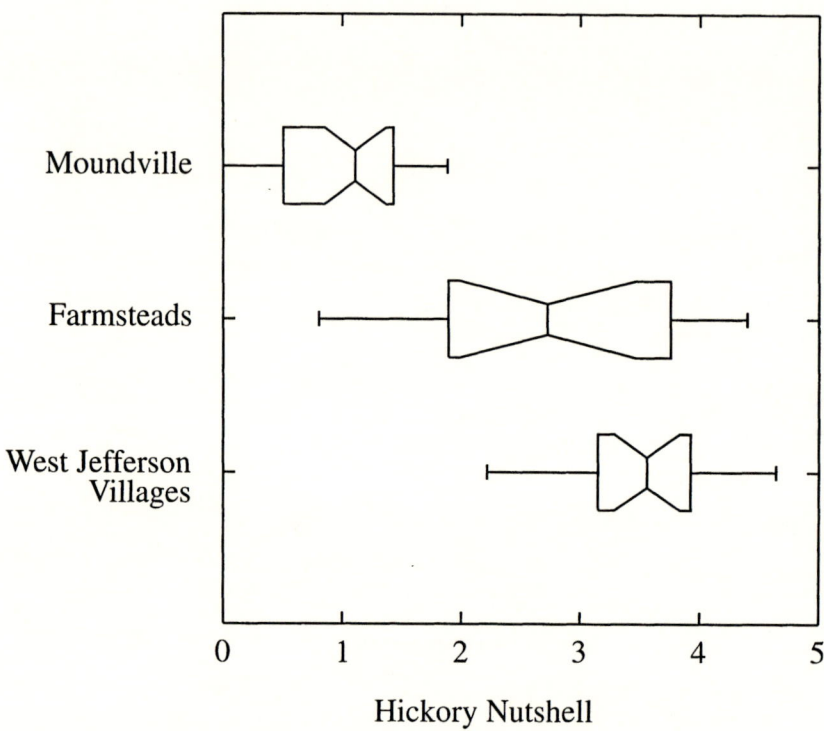

Figure 6.7. Boxplots comparing relative abundances of hickory nutshell at Moundville, the farmsteads, and West Jefferson villages. Values plotted are standardized counts reexpressed as natural logarithms.

If we allow that foods may have been processed and consumed at different locations, then we can offer the following interpretation for the patterns we have described. The relatively low quantities of food by-products at Moundville may be an indication that the residents of the farmsteads were sending provisions to the residents of the paramount center. That is, not all of the food produced or procured by the residents of the farmsteads was consumed by them. Rather some may have been partially processed to reduce its bulk and then sent to the subordinate and paramount centers. Several studies have indicated that preferred cuts of venison were provisioned from farmsteads or subordinate centers to the paramount center (Michals 1990; Scott 1981; Welch 1991). The larger quantities of corn cupules and nutshells at the farmsteads could indicate that plant foods were also being brought to the centers.

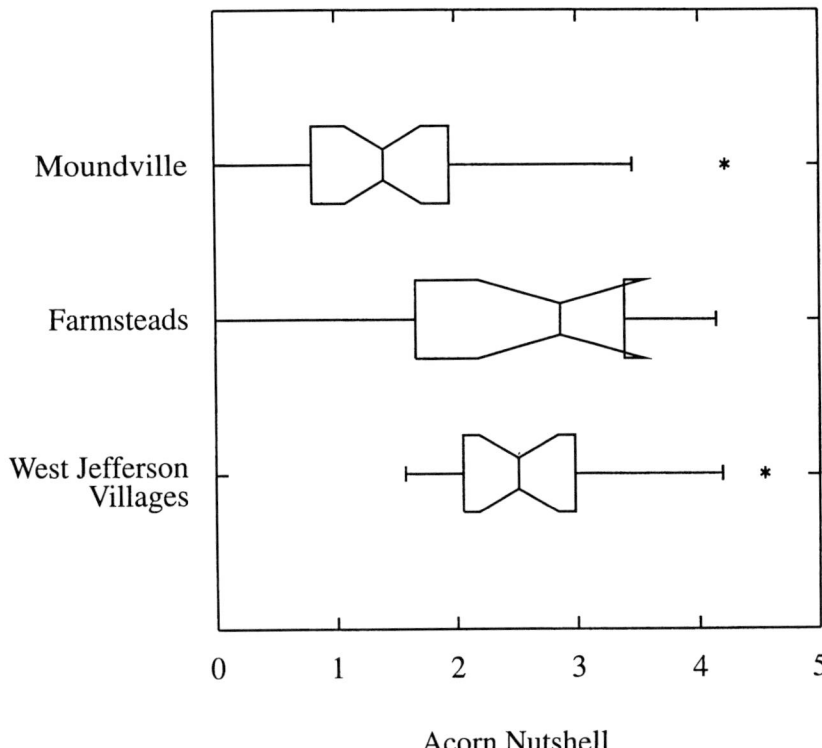

Acorn Nutshell

Figure 6.8. Boxplots comparing relative abundances of acorn nutshell at Moundville, the farmsteads, and West Jefferson villages. Values plotted are standardized counts reexpressed as natural logarithms.

If the residents of farmsteads provisioned the center, what were the economic and social relations that structured the flow of food? In a paper on the economic significance of Mississippian farmsteads Knight and Solis (1983) cited ethnohistoric accounts to address several issues relevant to this question. They noted that the links between people were not just those of elite to commoner or governors to governed. People were also bound by complex networks of kinship and obligation. This being the case, some food sent to the center may well have been channeled by, and perceived as part of, relations between kinfolk rather than being tribute paid by commoners to the elite.

Knight and Solis (1983) also called attention to the dual modes of crop production that existed among Native Americans in the Southeast. Families planted, cultivated, and harvested crops destined for their own larders. They also contributed labor to tending communal fields whose harvests were stored in granaries that were under the supervision and control of a chief.

Some communal fields were probably located at Moundville and its subsidiary centers where they were under the watchful eyes of the chiefs and their associates. It seems equally plausible, however, that some communal fields were located in the hinterlands. Farmsteads form loose aggregates in areas of good soils. Certainly it would have been expedient to locate communal fields where people could tend them without having to travel far or neglect their family fields. Such a practice would not only be labor efficient, it would also reduce the risk of poor communal harvests by spreading fields throughout the microenvironments in the valley.

How does this translate to the patterns we see in the plant data? Quite simply. If corn from hinterland communal fields was shelled before it was transported to the centers for storage or distribution by the chiefs, then quantities of corn cupules would be higher at farmsteads than in elite residences at the center.

Impact of the Moundville Polity on the Landscape

This book is organized around the theme of human-modified environments and agricultural evolution. We have described an evolving agricultural economy, but we have said little about its impact on the environment. Nevertheless, it should be obvious that the changes we have described did affect the landscape and how people related to it.

The 40 km portion of the Black Warrior Valley in which the Moundville polity was located is a complex mosaic of floodplain swamps, levees, and adjacent terraces surrounded by gently rolling hills. The best agricultural soils are on the levees and terraces; the terraces and uplands, when undisturbed, support mixed hardwood and pine forests in which hickory nuts and acorns abound (Scarry 1986).

At the beginning of the West Jefferson phase, we can envision people following a shifting-cultivation strategy. Villages and gardens would have been periodically relocated within a landscape composed of swamps, woodlands, and old village or garden sites in varying stages of succession. As old garden plots lay fallow, they would be invaded by numerous plants. Among the early colonizers would be maypops (*Passiflora incarnata*), blackberry (*Rubus* sp.), and other useful plants. Somewhat later, persimmon (*Diospyros virginiana*) and plum (*Prunus* sp.) trees might invade the old plots. These, and other fruit-bearing trees, might well have been left to produce when people recleared the plots for planting. Since nuts were most abundant in the rolling uplands, nut resources would have been little affected by garden clearing and other human activities on the floodplain.

As crop production increased in the late West Jefferson and Moundville I phases, fallow cycles may have been shortened. This would result in there being more land in early stages of succession. Bottomland soils, whose fertility was periodically renewed by flooding, might even have been continuously cropped. In either case, there would be a change in the relative abundance of wild fruits. Fruit from low-growing herbs and woody plants, such as blackberries and maypops, could have gained a foothold in idle fields or around field edges. In contrast, fields might have been replanted before shrub and tree fruits could become established.

The concentration of people at Moundville during the Moundville I phase may well have had a more profound impact on the landscape. Moundville is located on a high terrace that abuts the uplands. The innate fertility of the soil is very high, but the terrace is above the level of most, if not all, floods. If the land on the terrace was cropped continuously, or on a short-fallow regime, for the century or two that population was concentrated at Moundville, yields from household and communal fields probably declined. Under such circumstances, it is possible that people found it increasingly difficult both to supply their basic needs and to fill the chief's granaries with the harvests from the land around Moundville. Tribute collected as a tax on farmstead production or communal fields located in the hinterlands could have mitigated local shortages.

The concentration of people at Moundville and the construction of the civic-ceremonial center itself could have had impacts on the landscape beyond those created by the agricultural economy. Clearing fields would have had little effect on the availability of nuts since the uplands were nearby. However, when the population was concentrated at Moundville, per capita consumption of nuts may have decreased if large numbers of people tried to forage in the forests near Moundville.

The demand for wood for fuel and construction purposes must also have been high. Pines with their straight trunks and rot-resistant wood might have been in particular demand for building the palisade and other large public structures. The palisade was rebuilt at least three times (Scarry 1995; Vogel and Allan 1985). On the basis of the size and spacing of the palisade timbers, we estimate that a minimum of 10,000 logs would have been used each time the fortification was rebuilt. Even if the entire palisade was not rebuilt each time repairs were made, harvesting enough timbers for the palisade and other structures would have opened the forest canopy and perhaps altered the composition of the upland forests near Moundville.

In closing we would like to suggest that not all the changes in the landscape were a result of economic activities. Today, yaupon holly (*Ilex vomitoria*) is one of the dominant shrubs in the ravines and thickets at Moundville. Yaupon is native to the coastal areas of the Southeast, but the plant is found at a few locations well away from the coast. Fifty years ago the botanist Roland Harper (1944) wrote, "Away from the coast it is scattered so erratically as to suggest that it might have been planted by the Indians, who brewed a beverage from its leaves." The beverage to which he referred was the black drink that played a central role in many civic and ceremonial gatherings. It seems to us that Harper was likely right. Moundville's residents probably did indeed plant and tend the yaupon so that they could have ready access to the leaves of this ritually important plant.

CHAPTER 7

An Evolutionary Ecology Perspective on Diet Choice, Risk, and Plant Domestication

BRUCE WINTERHALDER AND CAROL GOLAND

$\mathbf{M}$ost models of plant domestication and the origins of agriculture assign causal primacy to one or more generalized, normative[1] variables (or "prime movers"; summary in Redding 1988:57–60). Examples are population, climate change affecting resource abundance or distribution, technological innovation, and energy-extraction efficiency. Typically these variables describe system-level properties and they are characterized by well-behaved mathematical averages. Population grows, climate shifts, technological proficiency improves, or energy use expands. Processes of change are removed from the daily decisions of individual ecological actors, and they are continuous and gradual. The agroecological and socioeconomic consequences cited in these models are the result of broad adaptive responses to steady changes in material conditions as the population moves from food gathering to food production.[2]

In this chapter we shift perspective and use evolutionary ecology models to examine the possible consequences for agricultural origins of localized decisions about resource selection. We approach the question of domestication through foraging models, beginning with a brief description of some key assumptions and concepts of that approach. We use the diet-choice model to show how resource selection decisions could bring foragers into contact with potential domesticates and how this might affect population density and subsistence risk. We turn to hunter-gatherers and show that sharing and regional exchange have evolved as highly effective responses to unpredictable day-to-day success in the food quest. We then show that the field dispersion common to nonindustrial farmers functions to reduce risk in the same manner as does sharing among hunter-gatherers.

The heart of the argument is more speculative. We try to envision the evolutionary transformation from foragers who buffer risk primarily by sharing to farmers who mitigate it by field dispersion. To provide empirical moorings to this exercise, we review the record of agricultural origins in eastern North America in terms of our scenario. Next, we briefly compare aspects of our approach with several other recent models. Our summary highlights some strengths of microecological analysis of prehistoric economies and their transformations (Winterhalder 1993).

Domestication as an Evolutionary Question

In the usual approach to domestication a change in the key variable (e.g., population growth) poses an adaptive opportunity or challenge (e.g., food stress). One or more of the properties of agriculture known to result from domestication (e.g., greater productivity per area) are posited as a response to the opportunity or as a solution to the problem. We believe that this manner of conceptualizing the issues has three liabilities that can be mitigated using evolutionary ecology models: (a) it engages an adaptationist-functionalist form of argument that is less powerful than a more direct selectionist approach; (b) it focuses attention on the particular variable (e.g., climate) to the neglect of its adaptively salient properties (e.g., unpredictable variability); and (c) it tends to focus on highly generalized variables, deflecting attention from the actual subsistence choices faced by foragers-becoming-farmers.

Our alternative perspective (a) adopts evolutionary ecology models, (b) encompasses nonnormative properties of the environmental variables affecting subsistence adaptations, and (c) locates the key processes in the immediate and localized decisions routine to the economic lives of the actors. Because of their importance, we state each of these points as a principle, followed by brief elaboration and the references containing the fuller justification.

PRINCIPLE 1: *Selection-based explanations are more powerful than functionalist (or the closely related adaptationist) explanations.*

In most accounts the domestication of plants or animals is explained as an adaptive response to an environmental challenge (e.g., resource shortages caused by overpopulation or climate change). The advantages of cultivating plants or husbanding animals (e.g., increased yield) thus account for the origins of the practice. In general, the form is that of a functionalist argument: a benefit of the feature is presumed to explain its origin. Although widespread in the biological and social sciences, functionalist analyses are fraught with logical and empirical pitfalls (Elster 1983). In particular, they usually are not subject to a causal theory explaining how benefits act to produce the trait.

Many of these pitfalls can be avoided by paying careful attention to theoretical or methodological underpinnings. In Elster's (1985) terms this means using aspects of methodological individualism to illuminate system microfoundations. In evolutionary ecology it means that analyses that begin with selection[3] and that attempt to deduce its consequences for behavior in specified environmental circumstances have proved more powerful than analyses that begin by trying to correlate possible functions or benefits with observed behaviors (Winterhalder and Smith 1992). Developed prehistoric agriculture may have had any number of advantages. However, the analysis that begins with the advantages attributed to mature agricultural systems may already have hopelessly obscured their possible causal effects during the process of domestication or the inception of agricultural production. In addition, the theoretical framework of evolutionary ecology places constraints on the types of benefits and beneficiaries that are plausible (Smith and Winterhalder 1992). For instance, it tells us to be skeptical of arguments that cite system stability as a benefit, large-scale aggregates (groups) of individuals as the exclusive beneficiaries, or future benefits as causal.

PRINCIPLE 2: *Analyses that include the nonnormative properties of an environmental variable are preferable to those that do not.*

Most accounts of domestication and agricultural origins rely on normative environmental analysis. The material conditions of life are set against an ecological background of shifting mean conditions. This approach has merit but underestimates the evolutionary importance of unpredictable or irregular temporal variation and spatial heterogeneity (Winterhalder 1980). It leads to an overemphasis on the source of the stress to the neglect of its properties. Description in terms of properties is especially important because it can be linked to general models of adaptive responses (e.g., Halstead and O'Shea 1989). Often the pattern of environmental variability—its frequency, duration, magnitude, spatial scale, and predictability—determines the qualities of a successful coping mechanism.

PRINCIPLE 3: *The more immediate a variable to the actual conditions and options experienced by an individual organism, the more likely it will be of causal importance in evolutionary analysis.*

The broad sweep of global climate change appears as if it invokes processes of analytical scope equal to the task of explaining so widespread and profound an economic change as agricultural origins. But prehistoric individuals did not face decisions in terms of climate change. The real stuff of evolutionary profundity lies in the local factors of an organism's experience: the constraints it faces, its actions, and the consequences of those actions. Immediate details have the greatest causal efficacy in the evolutionary explanation of adaptive design. Given analyses stated in terms of (*a*) changing climate, (*b*) changing habitat structure or distribution, (*c*) changing abundance of resources, or (*d*) changing prey selection or work efficiency, we should automatically prefer the last over the first. This will be true even if ecosystem processes link (*d*), (*c*), (*b*), and ultimately (*a*). The changes that we summarize under broad concepts such as *domestication* and the *Neolithic revolution* have their origin and form in the ecologically situated choices and actions of individuals.

Slobodkin (1973) makes a like point in a short essay critical of claims that ecosystems maximize some property such as ecological efficiency or constancy (we can add homeostasis, energy utilization, complexity, etc.). Slobodkin's reasoning turns on the differences between "extensive" and "intensive" variables. Extensive variables are those measures that summarize populationwide, interspecific (community level), or long-term (multigenerational) aspects of things biological. They refer to (emergent) properties unique to sets of organisms. Extensive variables such as ecosystem efficiency or complexity rarely find a causal role in evolutionary ecology models because they are not factors directly subject to the action of selection at the level of individuals. Intensive variables by contrast are those that characterize the behavior of an individual at a particular place and time. They refer to the situated properties of the organisms making up ecological sets; they potentially are subject to the direct action of selection. Examples include prey or habitat choice, time allocation, or reproductive tactics.

Slobodkin's injunction, "[T]he genesis of things which we believe to be the consequences of evolution . . . ought to be stateable in intensive, or local, variable terms" (Slobodkin 1973:299), is useful to us because it can guide first-order preferences among models of agricultural origins even if the empirical basis for such a choice is limited. It should immediately make us wary of any model of domestication or agricultural origins that relies on extensive variables. If empirical observations were to confirm that ecosystems (including agroecosystems) maximize efficiency, complexity, or such, we would take an interest in that fact but would attempt to explain it first as the outcome of interactions among intensive variables.

These three principles are related and reinforcing. For instance, Slobodkin's distinction between extensive and intensive variables (Principle 3) turns on the same theoretical and methodological guidance that allows us to discriminate in accepting evolutionary benefits and beneficiaries (Principle 1). An agricultural origins model that cites system-level energy maximization is suspect under both principles because it is difficult to reconcile with the workings of evolutionary processes. Likewise, to cite a shift in mean environmental conditions (Principle 2) is to state a mathematical abstraction at best weakly coercive relative to the fluctuating, patchy environment experienced by individuals.

Our commitment to an evolutionary perspective is complementary to that of Rindos (1984), who argues that domestication developed through processes of coevolution between human beings and the resources they exploited. As a consequence of harvesting certain species, hunter-gatherers became unintentional agents of dispersal, protection, pollination, and so forth. This altered selective pressures on the foragers and the plants, leading to behavioral and morphological changes, respectively. We believe that foraging theory can supply hypotheses on questions that the Rindos model neglects: What circumstances led human beings to select certain species for exploitation? What are the economic and population processes that accompany growing dependence on domesticates and cultivation?

Envisioning a Transformation

Foraging models are theoretically consistent with the principles outlined earlier (Smith and Winterhalder, eds. 1992), and they allow us to analyze the processes that brought human beings into regular contact with species that through coevolution became domesticates. The formulation that follows joins two other preliminary applications of evolutionary ecology models to this problem (Hawkes and O'Connell 1992; Layton and Foley 1992; Layton et al. 1991; Winterhalder and Goland 1993).

The Diet Breadth Model

Central to our analysis is the issue of resource choice. The diet breadth (resource selection) model is the oldest and best known model of foraging theory (MacArthur and Pianka 1966; Schoener 1974; Stephens and Krebs 1986). It sometimes is called an "encounter-contingent" model, because it draws a conceptual distinction between searching for an acceptable food item and the pursuing and handling of different food types. Search costs are a function of resource density (encounter rates with acceptable items); pursuit and handling costs are peculiar to each resource type. The forager searches for all resources simultaneously. Upon encounter with a potential dietary item he or she must make a contingent assessment: pursue the present resource or continue searching in hopes of locating a more attractive one to pursue.[4]

In practice resources are ranked by their decreasing net efficiency for the pursuit and handling phase of foraging (figure 7.1). The top-ranked resource always is in the diet. The model determines how far down the ranked list the "optimal" forager will go. The summary rule is this: add the next item if its pursuit and handling efficiency is greater than the overall foraging efficiency of the diet without it and, conversely, stop expanding the diet and ignore the first item for which the return on pursuit and handling is less than the average return for search, pursuit, and handling of higher-ranked items.

Several additional results are of special interest. First, items of high enough rank to be in the forager's diet should always be pursued if encountered, however rare they might be. Conversely, those with a ranking that puts them outside of the diet should never be pursued, no matter how common. Second, as the abundance of highly ranked items diminishes, overall foraging efficiency declines and the diet will expand stepwise to include items of lower rank. The changing marginal rate of foraging efficiency determines whether the item at the boundary of the diet moves into or out of the optimal set. Third, any change that increases the pursuit and handling efficiency of an unharvested resource

The following example demonstrates important features of the diet breadth model (see Winterhalder et al. 1988).

The table immediately below shows the characteristics of eight hypothetical resources. The items range in size (energy value) from APREY (comparable, say, to a small ungulate) to HPREY (comparable to a tuber). In addition to energy value, each resource type is described by its average pursuit and handling time, the forager's energy expenditure rate during pursuit and handling, and its density (a component of encounter rates).

Resource Characteristics

Resource Type	Energy Value (kcal)	Pursuit/ Handle Time (min)	Pursuit/ Handle Cost (kcal/min)	Resource Density (#/sq km)
APREY	60000	1864	6	0.8
BPREY	24000	836	6	2.6
CPREY	1380	235	6	8.0
DPREY	3600	95	6	6.4
EPREY	2800	174	6	12.0
FPREY	1500	67	6	40.0
GPREY	240	26	6	300.0
HPREY	90	7	6	500.0

Three further variables jointly determine how quickly and efficiently the forager can locate potential resources. The forager's search velocity and width affect his or her rate of encounter with resources of varying densities; the caloric cost of the searching has been set at 4 kcal/min, roughly the energy expenditure of a walking human.

Forager Characteristics

Forager's velocity of movement:	0.500 (km/hr)
Forager's search width:	35.0 (m)
Forager's search cost:	4.000 (kcal/min)

Figure 7.1. The diet breadth model.

Figure 7.1. *Continued*

Analysis 1 (below) shows how the diet breadth model treats this data. Resource types (column 1) are ranked (column 2) by their decreasing net return rates for pursuit and handling. Because rank is a net measure per unit time, it need not correlate directly with size of the resource type. Column 3 shows the overall foraging efficiency that would result from a diet breadth that begins with only the first item (CPREY, a diet breadth, or db, = 1) and successively adds additional resources through HPREY (db = 8).

At db = 1, the forager's efficiency is 965.3 kcal/hr. Although CPREY returns more than 3100 kcal/hr when located, it is rare, and searching for it exclusively is quite costly. Overall foraging efficiency can be increased to 1062.8 kcal/hr by adding DPREY to the diet. Though this is a less profitable item, the forager more than compensates by reduced search costs as more resource encounters produce acceptable items. The same is true of APREY and BPREY, but addition of the fifth item, FPREY, results in a decline in foraging efficiency from 1209.9 kcal/hr to 1160 kcal/hr. Expansion to items of yet lower rank causes further declines in overall foraging efficiency.

Analysis 1

Name	Rank (kcal/hr)	Full Density Foraging Efficiency (kcal/hr)	60% Density Foraging Efficiency (kcal/hr)
CPREY	3163.4	965.3*	602.5*
DPREY	1913.7	1062.8*	691.2*
APREY	1571.3	1165.1*	827.4*
BPREY	1362.5	1209.9*	925.9*
FPREY	983.3	1160.4	941.6*
EPREY	605.5	1079.6	882.4
HPREY	411.4	948.6	775.1
GPREY	193.8	719.1	579.3

The peak foraging efficiency at db = 4 establishes an optimal diet of species C, D, A, and B (all marked with an asterisk [*]). Even if encountered, FPREY (or any item of yet lower rank) returns less energy for the time spent pursuing and handling it than the forager can expect from exploiting only the top four resources. The cost of picking up these suboptimal items is seen in the declining foraging efficiency for a db > 4.

Figure 7.1. *Continued*

Column 4 shows how changing resource encounter rates can affect the optimal diet. In this instance, the densities of resource types A through D have been reduced by 40% (to 0.48, 1.56, 4.80, and 3.48/sq km, respectively). This increases search costs and lowers overall foraging efficiency so that FPREY enters the optimal diet. Its pursuit and handling efficiency (983.3 kcal/hr) is now greater than the foraging efficiency for a diet breadth of only four items (925.9 kcal/hr).

A suboptimal resource item can also move into the optimal diet through an increase in rank. In Analysis 2 (below) the pursuit and handling time of FPREY has been reduced by 20% (to 53.6 minutes). No other variables have been changed. FPREY's pursuit and handling efficiency increases from 983.3 kcal/hr to 1319.1 kcal/hr, enough to move it into the optimal diet. Note that overall foraging efficiencies are the same for diet breadths up to 4 items, but a new peak foraging efficiency (1229.9 kcal/hr) is established with FPREY in the diet at db = 5.

Analysis 2

Name	Rank (kcal/hr)	Full Density Foraging Efficiency (kcal/hr)
CPREY	3163.4	965.3★
DPREY	1913.7	1062.8★
APREY	1571.3	1165.1★
BPREY	1362.5	1209.9★
FPREY	1319.1	1229.9★
EPREY	605.5	1135.5
HPREY	411.4	989.1
GPREY	193.8	742.1

above the marginal foraging efficiency will move that item into the optimal set (see figure 7.1).

Diet Selection and Domestication

We begin our scenario by positing a resource (one, for simplicity, a plant we label TD, for transitional domesticate) that has a pursuit and handling efficiency that is somewhat too low to make it an element in the optimal diet. We presume a decrease, for any of a variety of reasons, in the abundance of more highly ranked species. This lowers the marginal efficiency of foraging sufficiently that resource TD enters the diet and becomes subject to coevolutionary pressures such as described by Rindos (1984). Following foraging theory, these pressures can be separated for analytical purposes into those that affect ranking and those that affect density. The former profitability changes might correspond to ease of harvesting or processing (e.g., size of a cereal grain); density changes might reflect growth conditions (e.g., inadvertent human effects such as clearing or soil enrichment). Although it can be ambiguous in practice, we will show that this analytical distinction between density and pursuit and handling profitability is an important one.

Figure 7.2 allows us to examine the possible outcomes of this coevolutionary relationship. We begin with the case in which TD is low in density. Its ranking (profitability) slowly increases with coevolution. When it first enters the diet (*cell 3;* low rank, low density) it has a limited impact on the subsistence economy. However profitable it may be, it is infrequently encountered and harvested. It makes up a small portion of the foods consumed. Its use has little effect on the exploitation of more highly ranked items. In fact, little about the foraging situation will change as TD coevolves to a position of low density and high rank. Neither the list of other species exploited (diet selection), nor the degree of their exploitation, nor the density of the foragers should change significantly. Thus, even if it is a species subject to some degree of human management, a low-density resource might be absorbed into a forager's diet without causing any striking changes in the existing hunter-gatherer economy. This situation corresponds to incidental domestication in Rindos's terms (1984:138–66). We will observe below that some premaize indigenous domesticates in eastern North America may have followed just this pattern.

Imagine, however, that the TD is very abundant (*cell 1;* low rank, high density). By foraging logic it will be ignored despite its density until a decline in overall foraging efficiency (or elevation of its rank) brings it within the optimal diet. Once in the diet, however, its economic impact will be relatively great. Although of limited profitability, its high yield will initiate forager population growth (see Winterhalder and Goland 1993). With a relatively dense human population sustained by a prolific if marginally profitable TD, high-ranking resources (those

Diet Choice, Risk, and Plant Domestication

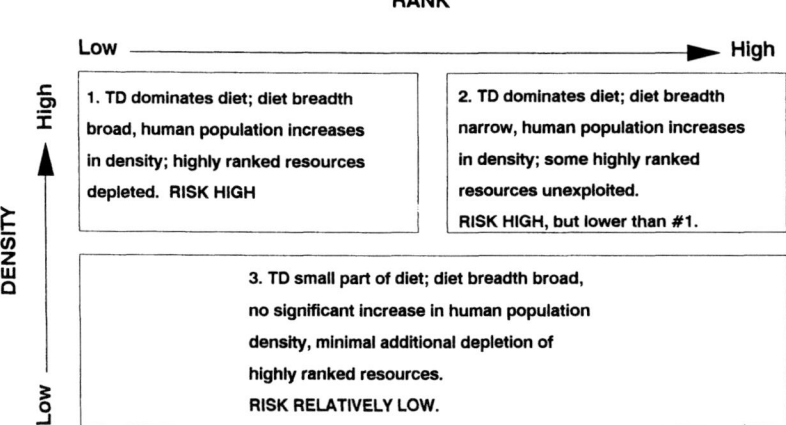

Figure 7.2. Model of plant domestication.

within the optimal diet) are likely to be overexploited and depleted. They are always worth capturing or harvesting when encountered. If some of these resource populations are locally extinguished the diet breadth may narrow. However, even if there is no local extinction, high-ranked items will become rare in the diet as their populations shrink. The foraging-becoming-agricultural population of cell 1 may find its subsistence regime dominated by the TD.[5] If the TD fluctuates stochastically in yield, the human population's risk will increase just as the option of compensating through the harvest (and perhaps sharing) of foraged foods is disappearing.

Cell 1 makes it evident that a minor change in the marginal efficiency of foraging can have quite significant economic ramifications, if the expanding diet happens to pick up an item of relatively high yield. Again referring to the Rindos (1984:138–66) terminology, this situation of relatively sudden dependence on a domesticate sets the stage for "specialized domestication."

In the last instance (*cell 2;* high rank, high density), we again posit that the profitability of the TD increases through coevolutionary processes. As in the case of cell 1, the diet will become dominated by the TD. However, the number of resources harvested will diminish as previously exploited foods are excluded from the optimal diet by the abundant and highly ranked TD. Risk will be high—the growing human population is increasingly dependent on the TD—but not so high as in cell 1. The items once harvested but now excluded from the diet constitute relatively profitable fallback foods. However, even at unexploited levels of population density, their yield might not long sustain a human group that has grown dense on the potential of a high-density TD in its most prolific years.

The overexploitation and depletion of foraged resources predicted in cells 1 and 2 may help to explain why domestication has strong elements of irreversibility (see also Layton et al. 1991:261) and why food production would spread once initiated in a locality. We note that it also is possible that rank and abundance both begin low and increase together, the coevolutionary direction moving from cell 3 to cell 2. In this intermediate scenario the impacts of a high-yield TD will develop much more gradually.

Evolutionary Ecology and Risk

The plant and animal resources of both foragers and farmers are subject to drought, flood, frost, epidemic, and other irregular calamities. In either system an adequate local food supply can be precarious. Episodes of scarcity are recurrent. Both foragers and farmers require adaptations that mitigate the effects of unpredictable subsistence shortfalls, and it is likely that these adaptations have quite dissimilar manifestations and consequences in the two economic modes. We suggest that there is much to be learned by considering how the structural parameters of risk and the effectiveness of alternate risk-buffering strategies change in the transition from food gathering to food production.

Because short-term yields are probabilistic, each subsistence choice can be characterized by an average reward and a variance. Risk reduction models demonstrate how foragers or farmers can select the mean/variance combination that maximizes the probability of survival by avoiding critical energy shortfalls. One of the clearest formulations of this approach is the Z-score model (Stephens 1990; Stephens and Charnov 1982). This model is based on a normal distribution, described in terms of an average harvest rate or income (m), the variability of income (s, measured in standard deviations), and a minimum requirement or critical threshold (R). The probability of falling below the minimum requirement is Z (the standard normal deviate), where $Z = (R - m)/s$. Risk is minimized as the value of Z decreases (figure 7.3).

Diet Choice, Risk, and Plant Domestication

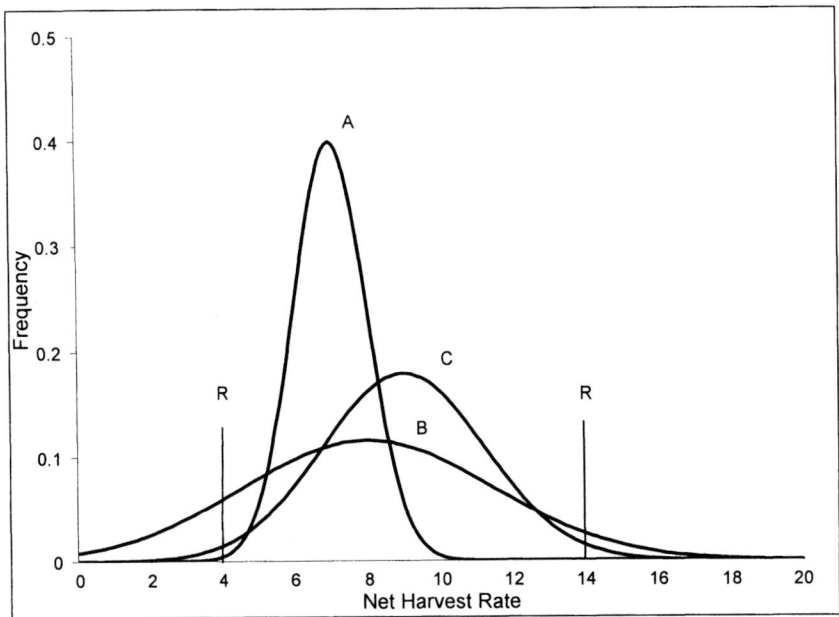

Figure 7.3. The Z-score model. Shown is the frequency distribution of net harvest rates for three resource options, with means and standard deviations as follows: (A) 7 and 1; (B) 8 and 3.5; and (C) 9 and 2.2. If $R = 4$, then the low-variance option (A) minimizes the chance of dropping below that value even though it also has the lowest average net harvest rate. If $R = 14$, then the high-variance option (B) offers the best (but not good) odds of success, even though it has a lower average return than option (C).

The model shows that the appropriate choice for enhancing probability of survival is dependent on the forager's minimum income requirement (R) and on the statistical properties (mean, m, and standard deviation, s) of the resource choices. When short-term requirements are greater than expected mean reward ($R > m$), the forager might do better to pursue more variant (high s or "less predictable") resource options, since they provide the greatest likelihood of fulfilling extraordinary requirements. Conversely, if needs are lower than expected rewards ($R < m$), as is more likely to be the case, probability of survival may be increased by selecting the less variant resource option (low s), even though it may on average yield less energy than another choice. As we describe below, the Z-score model summarizes the relevant considerations for both foraging and agricultural societies (Winterhalder 1990).

Foragers, Sharing, and Exchange

Hunter-gatherers face fluctuations in the encounter and successful capture of game on a day-to-day basis. To a lesser extent the same
kind of irregular variation will affect the availability and yield of plant
foods. In terms of the Z-score model, we expect a forager to select the
set of resources that, by a combination of high average return rate (m)
and low variance (s), minimizes the chances of falling below a critical
threshold (R). However, simulation modeling (Winterhalder 1986) has
shown that foragers may be able to reduce only modestly their risk of
food shortfalls by this means. Even the optimal selection of resources
to minimize risk may not efficaciously avoid it. For instance, in the
common situation in which $m > R$, adjustments to reduce variance
(thus the probability of a deleterious shortfall) typically also sacrifice
average foraging efficiency. Thus even as the offending left tail of the
food acquisition distribution (figure 7.3) gets pulled away from dan-
gerously low levels, the whole curve shifts toward lesser return rates.

Intraband sharing, or the regular pooling of food captured by individuals foraging separately, is a highly effective means of overcoming
this constraint (Winterhalder 1986). It requires only that the success
rates of those contributing to the pool are unsynchronized (that is, not
strongly positively correlated). Sharing groups with as few as six to
eight productive foragers gain the greater part of the marginal benefits
that result from this tactic. Given the ubiquity of environmental fluctuations in the lives of foragers, it is quite reasonable to view risk mini-
mization as an explanation for the sharing routinely observed in this
mode of production (Kaplan and Hill 1985).

If resources fail simultaneously over the whole range of a foraging
band, then the remedy available through pooling and spatial asynchrony
must be sought at a regional scale. Smith (1991) has explored the conditions that make exchange among bands located in different regions
an effective means of risk reduction. Most often this entails the temporary movement of foragers from areas of shortfall to those of greater
abundance. These residential shifts depend on a system of land tenure
that Smith terms *reciprocal access,* and although the costs and benefits can
be subtle, in general such arrangements appear to be quite efficacious.

Other risk-buffering tactics available to foragers appear to be less
effective. Carryover averaging (short-term storage) by a single forager
is mathematically equivalent to intraband sharing among a group of

foragers. However, it would take an individual practicing this tactic six to eight foraging intervals, some of them unsuccessful and hungry, to gain the same benefit achieved daily within a sharing group of six to eight individuals. In addition, carryover averaging entails processing costs and storage losses that do not affect intraband sharing. The ecological conditions under which long-term storage will be an effective counter to periodic scarcity are also fairly restricted for hunter-gatherers (Goland 1991). Periods of resource scarcity and high yield must alternate with regularity over long periods if storage is to become an important and routinized aspect of hunter-gatherer economies. In good years quite high yields must be attained, both to provision the population at a level that will permit extended residence at the storage locale and to provide the surplus for processing, storage, and storage losses. Because of such constraints, storage is infrequent among ethnographically documented hunter-gatherers.

Of the five options (resource selection, intraband sharing, regional movement, short-term carryover averaging, and long-term storage), sharing and regional movement appear to be the most effective for foragers. Empirical findings confirm that both act as effective risk-buffering strategies. Kaplan and Hill (1985), for instance, found that food sharing among Aché foragers occurs when (1) individuals experience fluctuations in yield, sometimes acquiring foods of higher energetic value than can be immediately utilized by an individual or family, and (2) returns of foraging among individuals are not highly (positively) correlated. Aché foragers most frequently share meat and honey, precisely the resources that are unpredictably and infrequently encountered and that provide more calories than can be consumed by an individual at one time. As a consequence of sharing, all Aché group members eat a diet that is more predictable than any of them would be able to capture on their own. Smith (1991) describes ethnographic cases of risk-buffering through reciprocal access. Prehistoric foragers presumably adopted similar mechanisms.

Food Producers and Field Dispersion

The same Z-score model that illuminates the situation of foragers has a direct analog among food producers: pooling of resources from spatially dispersed plots that are likely to have unsynchronized yields.

Field dispersion characterizes nonindustrial agriculture in much of the world. Agricultural economists and economic historians usually have depreciated this arrangement as wasteful and inefficient. They point to the costs of moving labor and materials among small, scattered parcels. The economic historian Donald McCloskey (1975, 1976) has explored instead the potential risk-reduction benefits of field dispersion. He uses the example of the open-field system found in the Medieval Midlands of England. Each year families planted wheat or barley in up to a dozen dispersed plots within the common fields of the manor. Other household fields were left in fallow. McCloskey demonstrates that given the localized vagaries of weather and pathogens, along with other spatially heterogeneous or "patchy" qualities of the agricultural landscape, a harvest that pooled yields from dispersed locations provided a more secure living. He calculates that compared with results from a single consolidated field, scattering doubled the probability of surviving twenty years without a subsistence calamity; it tripled that probability for a thirty-year interval. McCloskey's analysis shows that the greatest gains in risk reduction can be achieved by pooling the yields of about eight fields. His "safety first" model is a direct analog of the Z-score approach.

In the contemporary, high-altitude Andean communities of Cuyo Cuyo that we have studied (Goland 1992a, 1992b), families each year plant as many as two dozen scattered agricultural plots. These are dispersed across a topographically rugged landscape in which many of the agroecological factors affecting production (precipitation, edaphic conditions, pest infestations, etc.) are patchy and unpredictable. Yields from individual potato fields vary widely. If families were to rely on a single large plot rather than multiple smaller ones, the coefficient of variation of their production (standard deviation/mean $\times$ 100) would be about 75. As their holdings are broken into dispersed fields the coefficient of variation drops steadily. For three fields it averages 43; for six fields, 21. This is a striking benefit. We also have measured the costs of field dispersion that arise from added travel and transport. They reduce net productivity by 7%,[6] a noteworthy loss for a community on the margin of adequate production, but not an unduly large one.

With both the benefits and costs considered, frequency of disaster (which we define as failure to attain a critical threshold of calories) shows dramatic reduction as a household's fields are scattered. Of nineteen families studied, twelve reduced their probability of disaster to 0% by cultivating multiple fields. Under a hypothetical scenario of consolidated production these same families had probabilities of catastrophic shortfalls between 12.5% and 57.1% in the year sampled. Each of these families has minimum requirements (disaster thresholds) lower than the average potato yield for their fields. They illustrate the first half of the rule of thumb generalized from the Z-score model: if you can expect more than your requirement, choose the small variance subsistence option or, in this case, plant more scattered fields.[7]

A field that produces well one year may yield poorly the next. Cuyo Cuyo families cannot predict in advance of planting (even well into the agricultural season) how each of their fields will produce nor do they have the flexibility to disperse quickly or consolidate their landholdings. Given these constraints, an effective solution to subsistence risk is a scattered field system. Despite the added 7% cost, in a situation of marginal production and high risk like that experienced by Andean peasants, field dispersion dramatically reduces the odds of household production calamities.

The risks facing an agricultural household were presumably no less in prehistory than for contemporary farmers living in marginal environments. Nor, despite being situated in high mountain terrain, does the Peruvian case appear exceptional when compared with the degree of interannual subsistence variance and (intraannual) asynchrony documented for the Midlands of England. This raises the possibility of strong selective pressures for field dispersion or a comparable risk-reducing mechanism during the nascent stages of agricultural production.

Subsistence Risk During Domestication

We can now ask about risk as the various changes depicted in figure 7.2 take place. In the low-density cell (cell 3) incorporation of the TD into the diet causes no changes that would impede the risk-reducing effectiveness of sharing, exchange, or temporary, regional migration. Diet breadth remains broad and human population density presumably is low and relatively stable. The mean and variance of foraging efficiencies should change only modestly; exploitation pressure on harvested resource species does not jump dramatically. We propose that low-density TDs will function economically more or less like the undomesticated resource items in the repertoire of the hunter-gatherer.

However, in the high-density cases (cells 1 and 2) we must consider several new elements: human population density increases and humans grow increasingly dependent on one or a few species and thus more susceptible to their stochastic fluctuations. In addition, alternative food sources are either depleted by overexploitation (cell 1) or, even if unharvested (cell 2), perhaps less able to mitigate a food crisis afflicting a high-density human population. The type of pooling practiced by foragers will have diminished effectiveness, producing selection for alternatives.

The changes that will occur are constrained by two structural elements that also will be shifting in this transition. One is temporal and the other spatial. First, although there can be strongly seasonal elements to the food quest of foragers, typically their production interval is measured in days. From start to conclusion in success or failure, several complete production episodes (hunting trips, gathering expeditions) can take place before hunger would begin to debilitate the forager. The temporal scale of consumption roughly matches that of the production cycle. However, as annual crops begin to dominate the diet of temperate-zone agriculturalists, the production interval lengthens to a year. Months pass between planting and harvest; only near or at harvest is the success or failure of the productive effort fully evident. Months more may pass before another planting can be initiated. The temporal scale of production for food producers in a seasonal environment is much greater than that of consumption.

This difference in production interval has at least two consequences for the present argument. First, long-term temporal averaging will not be as effective for food producers simply because they cannot survive through several failed production intervals (whereas foragers, with their short production interval, might survive). Larger processing costs and greater storage losses that accompany the longer period (months or years versus days) needed for temporal averaging among agriculturalists further reduce the attractiveness of this option. Second, any pooling strategy is vulnerable to the impact of cheaters, the laggards who slack off on their own efforts but who receive a full share of the combined effort. Hunter-gatherer groups certainly face this problem, but for foragers the relatively short duration of the production interval allows for quick assessment, sanction, and correction before imbalances grow too large. Detection of a free rider will be significantly more difficult for farmers, who might inadvertently subsidize a shirking neighbor for years before being able to detect his or her deficiencies. Likewise, correction through social control may be more difficult when the production interval is measured in years.

The second structural element that differs in the two cases is the spatial dispersion opportunities. The one or two active foragers of a hunter-gatherer domestic unit are constrained in their ability to take advantage of unpredictable and spatially uncorrelated resource opportunities. This is because the foraging mode of production is both immediate and active, requiring the presence "on the spot" of the hunter-gatherer. The forager realizes the resource possibilities of a location only by being there at the right time. Effective spatial averaging requires intraband pooling. Farmers do not face this constraint. By dispersing fields a lone farmer can gain all of the advantages of spatial pooling from multiple environmental sites. Production at a particular locality does not require the individual's presence, except for periodic attentions of plant cultivation as the growing season progresses, and these can be staggered. In short, an individual forager cannot be hunting and gathering in six places at once; the farmer, however, can plant, manage, and harvest fields that are in six dispersed locations. Foragers are constrained to pool on an intraband basis; farmers have the option of adopting an interhousehold (perhaps at the village or kin level) or an intrahousehold approach.

Despite this option, we propose that populations increasingly dependent on a few high-yield domesticates will face selection pressures to move risk management within the household. Temporal averaging (beyond routine, seasonal storage) is costly because it implies a large surplus above annual needs, high processing costs, and potential losses in food that must be stored for several years. The free-rider problem is exacerbated by the long production interval and denser population, which thereby reduces the attractiveness of interhousehold pooling. We hypothesize that in the circumstances of early agricultural production the costs of intrahousehold field dispersion (say, 7% to 10% of net production) were less than the free-rider costs attached to continued interhousehold pooling. We envision the transition accompanying the origins of agriculture to be one in which interhousehold sharing eventually was supplanted by intrahousehold field scattering (supplemented initially or eventually by regional trade). We propose that community controls and social relations among early farming households were aimed primarily at organizing dispersed production opportunities rather than interhousehold distribution. We again draw on our Andean research to provide an example.

Field scattering in the Andes is associated with sectorial fallowing systems (Orlove and Godoy 1986). Households manage the production process through community-level coordination of planting schedules and rotations. Agreement on the initiation of planting dates ensures that animals will be removed from the zones slated for cropping, thereby reducing the probability that they will damage crops. Coordination of crop rotations also reduces the threat of pest and disease infestations and soil depletion. Community-level controls associated with the sectorial system regulate rights in the control, use, and transfer of property in land. They protect isolated fields, which are vulnerable to theft and other depredations. We propose that early agricultural communities are under strong selection pressures to develop similar coordination mechanisms, especially if households claim and invest in multiple, dispersed fields.

Social relations among households are key in the allocation of labor in agricultural communities. We earlier noted that fields do not require continuous presence, supervision, or input. Nonetheless, production is constrained by peak periods of labor demand (planting, harvesting) at key points in the agricultural cycle. In the Andes, shortages of workers within the household are met by interhousehold labor exchanges that draw on mutual assistance (*ayni*). These relations provide the fabric for community functioning and reproduction. In contrast to the case for hunter-gatherers, social relations in these agricultural communities are built on seasonal reciprocities of labor rather than on continuous reciprocities of food.

We predict that field scattering built around sectorial rotation systems or their equivalent would have characterized early agricultural communities, with these or like implications for social structure and interhousehold political life. The regional exchange mechanisms emphasized by archaeologists (see Rindos 1989) may have augmented these mechanisms or may have appeared somewhat later, coincident with the development of regional sociopolitical organization. Yet later, centralized political administration (e.g., the Inca State) may have complemented or supplanted these household-level means of insurance. The specific hypothesis that we have proposed is set within a more general and somewhat separable claim, to the effect that changing patterns and sources of subsistence risk may have been more important to socioecological adaptation during the origins of agriculture than were changing production efficiencies.

Discussion

In the following discussion we set our proposal in the context of plant domestication and agricultural origins in eastern North America, compare it briefly with three other recent models of plant domestication, and comment critically on the widespread use of energy efficiency to frame questions about agricultural origins.

Plant Domestication in Eastern North America

Although opinion has shifted several times during the past century, since the mid to late 1980s the consensus has held that eastern North America was an independent center of plant domestication. The earliest possible domesticates include bottle gourd (*Lagenaria siceraria*) and gourd/squash (*Cucurbita pepo*). The former has been identified in archaeological contexts as early as 7300 B.P. (Doran et al. 1990); its status as a domesticate at this early date has not yet been clearly determined. A stronger case can be made for domesticated *Cucurbita pepo*. Fragments of *Cucurbita* have been recovered from archaeological contexts as early as 7000 B.P. (Smith 1992b). Some researchers (e.g., Asch and Asch 1985b) believe that these early occurrences represent an introduced tropical cultigen; others (e.g., Decker 1988; Fritz 1990 and this volume; Smith et al. 1992; Watson 1985) feel that these represent a native wild gourd, independently domesticated after 3000 B.P. in the eastern United States. Support for the great antiquity of this taxon north of Mexico is provided by the occurrence of *Cucurbita pepo* seeds from preoccupation levels associated with Pleistocene megafauna at the Page-Ladson site in northern Florida (Newsom et al. 1993). The domesticated status of three other species is more secure. These are the oily-seeded sumpweed (*Iva annua:* see Asch and Asch 1978, 1985b; Ford 1985; Yarnell 1972, 1978, 1986) and sunflower (*Helianthus annuus:* see Ford 1985; Yarnell 1978, 1986) and the starchy-seeded chenopod (*Chenopodium berlandieri:* see Fritz and Smith 1988; Smith 1984, 1985a, 1987a; Smith and Cowan 1987). The evidence for domestication of sumpweed and sunflower rests on increased achene size (Asch and Asch 1985b; Smith 1989; Yarnell 1978). Domesticated *Chenopodium* is recognized by reduction in the thickness of the seed coat and associated morphological changes (Fritz and Smith 1988; Smith 1984, 1985a, 1989; Smith and Cowan 1987). These plants were regularly harvested in small quantities in their wild form perhaps as early as 8000–7000 B.P. By 4000–3000 B.P. they had been brought into cultivation and had undergone morphological changes indicative of domestication. Throughout this long period of use and cultivation they appear to have made only a modest contribution to the diet (Gremillion 1993c, 1994; Smith 1989).

Several other native plants were also regular components of the prehistoric diet, but it is less certain that they were domesticates. They are identified as possible crops because they are recovered from archaeological contexts well beyond their natural geographic ranges or in quantities great enough to suggest harvest of managed stands. Cases have been made on archaeological, ecological, and/or geographical grounds for prehistoric cultivation of maygrass (*Phalaris caroliniana*) (Cowan 1978a), giant ragweed (*Ambrosia trifida*) (Fritz 1990), little barley (*Hordeum pusillum*) (Watson 1989), and erect knotweed (*Polygonum erectum*) (Asch and Asch 1985b; Fritz 1990). As yet, however, no clear-cut morphological indicators of domestication have been identified for these plants.

Several recent hypotheses about plant domestication in eastern North America (e.g., Smith 1987b, 1992c; Watson 1989) have incorporated elements of Anderson's (1952) "dump heap" theory. Smith (1987b, 1989) subsumes this idea within a coevolutionary framework like that proposed by Rindos (1984), with climate change as the initial trigger. He argues that Hypsithermal warming (from 8000 to 4000 B.P.) led to changes in stream flow and alluvial deposition patterns that resulted in the formation of more productive aquatic habitats in the midcontinental area. Because of this enrichment of river valley resources, floodplain locales were visited more frequently and occupied for longer times than during earlier periods. The net effect of these two trends was to foster the development of anthropogenic habitats (which Smith terms *domestilocalities*).

Smith believes that plant cultivation and domestication occurred in these contexts. In this process, the significant properties of the domestilocalities included increased sunlight, enhanced soil fertility, continual soil disturbance, and the constant introduction of seeds. Weedy colonizers invading domestilocalities were subjected to selection pressures that produced two changes: seed production was increased and seed dormancy decreased. Both improve a plant's competitiveness in the disturbed habitat (see also King 1987). Neither of these changes requires human intervention but each makes a plant more attractive because they enhance its economic potential.

According to Smith, once people began to utilize a plant they became agents of intervention in its life cycle. At first inadvertently, for example, through creation of disturbed habitats and dispersal of seeds, human beings encouraged the growth of plants with economic potential. Eventually interventions became intentional, leading to husbanding activities such as weeding. This reduced interspecies competition and may have increased the stand of the quasicultigen plant. Planting seed was the next significant step. Harvesting by human beings of planted areas and replanting of harvested seeds automatically selected for increased seed retention and more compact, terminalized seed heads, along with uniform seed maturation, which led to greater harvest yields. Seed bed competition selected for reduced dormancy, more rapid growth, and increased seed size. Smith believes this accounts for the morphological changes observed on specimens of sumpweed, chenopod, and sunflower dating from the period 4000–3000 B.P.

In the last half of the third millennium B.P., economies based on significant food production emerged in the Midcontinent. Though not all were utilized in every region, the main plant foods were the domesticates gourd/squash, chenopod, sunflower, and sumpweed; crops such as knotweed, maygrass, giant ragweed, amaranth, and little barley were also economically important in many localities. But despite their long period of cultivation and use, the native domesticates never dominated the diet (Ford 1981; Watson 1989). Throughout the Archaic and subsequent Woodland periods, up to about A.D. 1000 (late Late Woodland), the inhabitants of the Midcontinent essentially maintained a foraging economy, relying on wild foods, especially nuts, and terrestrial or aquatic fauna. Thus native crops did not become significant food sources until long after their domestication. Smith (1989) suggests that these garden crops provided an important storable resource for the critical late winter–early spring period when alternate food supplies were low.

There are scattered identifications of maize in the archaeological record beginning in the Middle Woodland period. The earliest occurrences are dated to 2017 ± 50 and 2077 ± 70 B.P. at the Holding site in the American Bottom of Illinois (Riley et al. 1994), 1775 ± 100 B.P. at the Icehouse Bottom site in Tennessee (Chapman and Crites 1987), and

1730 ± 85 and 1720 ± 105 B.P. at the Edwin Harness site in Ohio (Ford 1987). Although maize is present in archaeobotanical assemblages infrequently and in small quantities for roughly 600 years (Smith 1989; Yarnell 1993; Keegan and Butler 1987), in collections dating to about 800 B.P. its archaeological visibility begins to increase dramatically. By A.D. 1150 maize-centered field agriculture emerged across much of the Eastern Woodlands, sometimes accompanied by beans and squash, though the importance of these two crops was variable throughout the region (Smith 1992c; Watson 1989). Skeletal studies of dietary patterns parallel the archaeobotanical record of maize (Fritz 1990; Smith 1989; Wagner 1986). For example, in west central Illinois carbon isotope data indicate that maize was present for several centuries before it became a substantial component of the diet (Buikstra et al. 1987). Thus maize may have been an experimental or minor garden crop during the initial 400- to 600-year period of its use, but after ca. 1000 B.P. (depending on the region) maize production was intensified, field agriculture became more extensive, and maize came to dominate the food production system (Fritz 1990; Smith 1992c).

Considering this evidence on prehistoric domestication in light of our model will shed light on the interrelationships of resource density, rank, diet breadth, and risk. We leave aside consideration of early squash and bottle gourd, since these may have been used as containers rather than as dietary staples (however, see Cowan, this volume). The evidence demonstrates that sumpweed, sunflower, and chenopod exhibited morphological indications of domestication by 4000–3000 B.P. We propose that after they initially entered the diet they were exposed to the co-evolutionary forces responsible for these changes. We see two possible and nonexclusive pathways to this result. First, following Smith, changes in stream flows and landscape development during the Hypsithermal induced longer-term occupation of localities. Localized depletion of the resources surrounding the occupation sites would result. This depletion, which would require foragers either to travel farther to collect the same quantity of food or to collect foods of lesser caloric value locally, decreased overall foraging efficiency. Diet expanded as foragers incorporated lower-rank resources previously ignored, including the earliest native domesticates. A second scenario posits that the same alteration, increased duration of occupation at bottomland sites, created more disturbed habitats in which the candidate cultigens could invade and thrive, as Smith's "floodplain weed" theory postulates. If their growth were favored in these areas, then by virtue of their more proximate location and/or more dense growth, their profitability might rise enough so that they would enter the optimal diet,[8] initiating the process of domestication under coevolutionary pressures.

By either of these mechanisms, a change in settlement pattern might alter the diet to include early domesticates. However, our model suggests other alternatives (figure 7.4) not requiring alterations of scheduling around the postulated domestilocalities. For instance, a simple reduction in the number of high-ranked resources, whether because of climate, overexploitation, or other causes, would be sufficient to bring foragers into contact with potential domesticates.

With reference to figure 7.2, we posit that the initial native domesticates were low-density resources that remained in the lower cell throughout the long period of their prehistoric use. Even if exploited and grown for several thousand years as domesticates, we expect their impact on diet and economic activities to be limited. The prehistoric record indicates that inhabitants of the Midcontinent maintained a broad-spectrum economy that included hunting and gathering while also cultivating and domesticating plant foods that made a relatively modest contribution to subsistence. We would also expect relatively little change in population density during this long period, a prediction more difficult to assess with empirical evidence. Prehistoric population trends are difficult to identify and quantify and even when recognized are often specific to extremely localized regions.

In contrast, maize was an introduction that eventually achieved the qualities of a highly ranked and potentially dense resource (the upper half of figure 7.2). It came to dominate the diet as human populations increased in density. The prehistoric record indicates that the proportion of the diet composed of maize increased relatively rapidly, beginning about A.D. 800. This would represent a quick shift from cell 3 to cell 1 or 2. Changes in the technology of food production or characteristics of the maize variety itself could be responsible for this development.

Depending on its rank, we can also make specific predictions about the impact that expanding maize dependence had on the exploitation of other food resources. If maize was a low-ranking resource, then diet breadth should remain broad, but other nondomesticated resources of higher rank should show evidence of overexploitation, depletion, and reduced absolute contributions to the diet (cell 1). If maize was (or eventually became) a high-ranking resource, then diet breadth should

---

Possible Causes of Increased Foraging Efficiency

1. Increase in the density of high-ranked food items (those in the diet), thus the forager's encounter rate, due to habitat improvement, game population cycles, release from over-exploitation, etc.
2. Reduced search costs, perhaps due to decreased energy expenditure in movement.
3. Changes in resource distribution (e.g., resources become more spatially aggregated).
4. Increase of pursuit and handling efficiencies of items in the diet (see below).

Possible Causes of Increased Rank of a Food Item (i.e., increased efficiency in pursuit and handling)

1. Improved transportation in pursuit (e.g., snowshoes, horses, mechanized transport).
2. Improved technology of harvest (e.g., traps, firearms for mobile prey, sickles for plants).
3. Increased capacity for transporting produce (e.g., improved or lighter-weight containers, baskets, pack animals).
4. Improved methods of food processing
    a. More efficient tools for cutting, cracking, grinding, etc. (e.g., mortar and pestle for removing glumes).
    b. Better fuels (e.g., hotter burning firewood species).
    c. Improved technology in heat transfer in cooking (e.g., pottery rather than stone boiling, thin-walled pottery).
5. More effective storage methods (e.g., those that reduce storage loss) or storage facilities that are more efficiently constructed.
6. Morphological changes to the resource increasing its profitability (e.g., tougher rachis so grains hold together at maturity; reduction in toughness of glumes reduces threshing labor, larger fruits).

Figure 7.4. Possible causes of increased foraging efficiency.

narrow as items of lower rank are dropped (and their populations recover from exploitation).

The evidence to test these predictions is inconclusive at present. For example, Fort Ancient populations in the Ohio River drainage appear to have relied heavily on maize to the exclusion of most of the native cultigens (Wagner 1987). In contrast, most archaeobotanical assemblages from other maize-dominated sites demonstrate continued if reduced reliance on indigenous crops and wild species of plants and animals (Smith 1992c). Rindos (1989:31–32) notes that introduction of the new maize crop did not exclude cultivation of the starchy and oily seeds. In our terms, this suggests that maize entered the diet as a resource of low rank (cell 1) relative to the seeds, which by then had presumably moved toward the right in cell 3.

Our model leads us to believe that a population's exposure to risk will be significantly altered in the course of the coevolutionary processes of plant domestication. Risk will increase if a single high-density resource dominates the diet (cells 1 and 2). Maize-dominated agriculture may be a prime example (Rindos 1989). Risk will be especially high under conditions described in cell 1, in which a dominant high-density/low-ranked resource is accompanied by broad diet breadth, population increase, and depletion of highly ranked resources. Early farming peoples such as Mississippian populations may have faced greater risk at the same time that they had fewer strategies for buffering shortfalls, on account of population increases, reliance on fewer crops, and a diminished array of wild fallback resources because of environmental deterioration resulting from overexploitation and land clearing (Armelagos et al. 1991; Ford 1977; Rindos 1989; see also Scarry and Steponaitis, this volume).

O'Shea (1989) analyzes the ethnohistoric record for contrasts between risk-buffering mechanisms of the Plains-dwelling Pawnee and the Huron of the Upper Great Lakes. He focuses on the use of wild resources to buffer shortfalls in agricultural production. To be an effective buffer, a nondomesticate must (1) occur in dense patches so that it can be intensively exploited, (2) be storable, (3) exhibit interannual fluctuations in availability that are independent of those affecting agricultural production, and (4) have a pattern of intraannual availability that does not compete with agricultural pursuits and a periodicity capable of buffering seasonal agricultural shortages.

Pawnee subsistence, based on maize agriculture, was vulnerable to droughts and grasshopper infestations. Buffering strategies included the use of scattered fields, interannual storage, and diversification (the regular planting of maize varieties and the use of fallback wild foods in times of emergency). The Pawnee also emphasized social obligations for sharing within the village and for intercommunity exchange. According to O'Shea, the most important risk-buffering strategy was the large mammal hunt, principally of bison. Communal hunts were scheduled at times when agricultural labor was not needed, and they regularly provided a storable resource for the winter and early spring months of scarcity. During years when the harvest was poor, the hunts were intensified and could provide an enormous quantity of meat to be dried and stored.

Like the Pawnee, the Huron also depended principally on maize agriculture. Large village fields of corn were supplemented by other domestic crops and wild foods. The principal threats to agricultural production were frosts and drought. Buffering strategies included storage, a diverse subsistence base of crops along with wild plant and animal foods, and exchange between Huron villages and between the Huron and other tribes, but apparently did not include field dispersion. Despite these means, the threat of starvation was real: in the years between 1628 and 1650 severe crop failure was reported once every four or five years. Periodic hunger was a result in part of the lack of appropriate wild resources. The only fallback wild food available to the Huron that fulfills the requirements outlined by O'Shea (1989) was anadromous fish. In normal years the smoked fish from the autumn fishing could see the population through to the spring. In poor agricultural years this resource was insufficient to make up the agricultural shortfall. The amount available for direct consumption or use in exchange was limited by the requirements for processing and transport. Villages were not located near fishing grounds, and the small task groups sent to exploit the fish runs were unable to process and bring back large quantities.

These two examples highlight several important aspects of risk buffering in the midcontinental region. First, field scattering played a role in risk management where clearing was relatively easy (the Plains) and was not used where clearing exacted heavy labor costs and where high-quality soils were patchily distributed (the forested Great Lakes). Second, the Pawnee amplified the benefits of field scattering by pooling resources among production units. By contrast, the Huron planted in large village fields and so the harvests of individual production units would have been essentially identical and intravillage sharing of agricultural products would add little as a measure of security. Third, the ability of a wild resource to buffer agricultural risk depends on the degree to which it can be intensively harvested, processed, and stored. Although, according to O'Shea, the pattern of availability for anadromous fish is independent of the factors responsible for interannual and intraannual variability in agricultural harvests, the Huron's ability to utilize this resource to buffer agricultural shortfalls was limited by their capacity to harvest and store it in sufficient quantities.

Finally, the buffering capacity of wild resources depends in part on their potential yield relative to the human population. By late prehistoric times in the Great Lakes region the deer population had been significantly depleted and was thus unable to serve as a fallback resource. In contrast, the success of risk buffering among the Pawnee can be attributed at least in part to low human density relative to the (bison) resource base. As population increases, whether because of dietary improvement (as postulated in our model) or because of other factors (population aggregation for defensive purposes, as in the Huron case), the ability of wild resources to buffer the population from agricultural shortfalls can be diminished. Moreover, low population density in the Plains allowed communities to maintain autonomous territories that included hunting grounds. This territoriality, absent in the Great Lakes, allowed Plains villages to shift adaptive postures easily in order to emphasize hunting rather than agriculture. Low population density is thus also correlated with greater flexibility in responding to production crises.

O'Shea's ethnohistoric reconstruction indicates the importance of risk buffering in the agricultural economy of two early historic-period agricultural groups. The risk of agricultural production is apparent at earlier periods as well, though the evidence is less direct. Throughout the world a sharp decrease in the health of early farmers relative to that of hunter-gatherer forebears is fairly well established;[9] an increase in the frequency of episodic nutritional stress is less securely documented, but evidence is suggestive (Armelagos 1990; Cassidy 1984; Cohen and Armelagos 1984; Cook 1984; Goodman and Armelagos 1985; Goodman et al. 1984; Larsen 1984; Perzigian et al. 1984). For example, studies of dental enamel pathologic conditions (linear enamel hypoplasias and enamel microdefects) indicate more frequent and/or more severe episodes of stress in early farming populations compared with the hunter-gatherers who preceded them. We suggest that this apparent jump in sporadic food crises corresponds to the period in which high-density domesticates had significantly elevated subsistence risk (cells 1 and 2 in figure 7.2) but before effective risk-management practices had developed. We would expect such a lag because of the sociopolitical complexity of shifting to intrahousehold field dispersion or some other means of risk management.

Related Studies

From the extensive general literature on agricultural origins, we have selected three recent models for brief discussion. We focus on a few salient points of each model in order to sharpen comparison with our own approach.

Flannery and Reynolds (Flannery, ed. 1986:435–507) have designed a complex, semirealistic computer simulation that attempts to replicate the resource collection (scheduling) choices faced by a small microband of transitional foragers at Guilá Naquitz, a prehistoric cave site in the Valley of Oaxaca, Mexico. The model incorporates quantified data on prehistoric resources, their nutritional properties, and their spatial distribution. It allows resources to fluctuate through irregular sequences of wet, normal, and dry years. Members of the hypothetical microband periodically compare their resource-acquisition success (protein-calorie yield relative to effort) with that attained in earlier time units and adjust both the foraging tactics used (the resource-collecting schedule) and their more general rules for appraising success and choosing tactics (decision-making policies). The simulation begins with a randomly selected schedule; as it runs it traces the growing effectiveness of resource-harvesting decisions until a stable plateau is reached. At this point, the simulated band members harvest plant resources in relative proportions nearly the same as those actually documented from the excavated cave deposits.

The Flannery-Reynolds (hereafter FR) model emphasizes the information (knowledge) gained by foragers through trial-and-error learning. Although its authors disavow the term *optimization,* the FR simulation mimics a process of foraging optimization just like that which might be guided by selection: "The group is not instructed to 'optimize' or to 'maximize' any particular variable. They are merely told to make small changes in their foraging strategy each time step, to remember how well each strategy did, and to improve through time by repeating more successful strategies and disdaining less successful ones" (Flannery 1986:501). The second sentence of this citation describes a model that in fact does instruct the microband to optimize its foraging efficiency in precisely the constrained sense used by evolutionary ecology models.[10] There is one operational difference: the FR model replicates a process of optimization whereas an optimal foraging analysis usually assumes that the process will occur and attempts instead to predict the outcome. This difference aside, our observation that the FR model incorporates an optimization premise is of more than semantic importance because it removes a false distinction that would impede fruitful comparison of the two approaches.

Experiments with the FR simulation show that the rate of improvement in preagricultural scheduling strategies is greatest when variability is present (unpredictable sequences of wet and dry years). By contrast, in their model of incipient agriculture, gradual changes in average climate (slowly increasing humidity or aridity) or population density did not speed the rate at which cultivars were adopted into the plant utilization schedule. In effect, the empirical success of the FR model appears to depend on its incorporating stochastic environmental fluctuations. This independently confirms the importance of Principle 2. Redding (1988:71; discussed below) makes a similar point: "fluctuations have the potential for being a very important factor in the evolution of human subsistence."

Full comparison of the assumptions, formulation, and predictions of the FR approach and evolutionary ecology must await another opportunity and fuller development of the latter. Though the FR simulation is quite unlike anything developed so far by foraging theorists, the premises and form of this simulation are concordant with the principles cited earlier.

A second recent model of domestication and agricultural origins is that of Redding (1988). Redding states premises much like those of evolutionary ecology: he proposes a focus on the selective pressures affecting subsistence choices, is skeptical of group-based functionalist explanations, and gives unpredictable interannual fluctuations in food supplies a key role in his model. He views the transition from foraging to food production as an evolutionary one of shifting subsistence tactics and strategies.

Redding's model mixes population growth and unpredictable resource abundance. He proposes that risk avoidance becomes important as foragers first approach the carrying capacity of their environment because resource fluctuations matter more under conditions of general scarcity. Redding postulates that foragers pass through a sequence of responses. Migration to a more favorable zone is the least expensive and thus the first buffering tactic to be used. Its success elevates the effective carrying capacity of the environment. However, with growth the population again comes under selective pressure of resource fluctuations. Diversification is adopted next. Population growth resumes and the cycle repeats with storage being the third response of choice and food production being the final risk-avoidance tactic adopted. In effect, food production is the last-introduced and most costly of a sequential set of risk-mitigating tactics; it is initiated to supplement gathering when all other options for coping with periods of shortfall have been exhausted. Redding states, "Without population pressure and unpredictable, severe depressions in the subsistence base, individuals that invest labor in planting a wild resource or caring for captured animals would be wasting their time; there would be no selective advantage to planting and herding" (Redding 1988:80). For Redding, plant cultivation is the answer to the short-term risks of foraging.

While the Redding study shares premises with evolutionary ecology, in making them operational it appears rather more like the older functionalist-adaptationist models. It is a refinement of the population-as-prime-mover argument: problems of hunter-gatherer population pressure are solved with agriculture. However, in Redding's view, the problem is not one of increasingly severe chronic food shortages. Rather it is one of resource fluctuations that grow in severity as the population approaches carrying capacity. Of possible solutions, agriculture is the last taken up in a sequence of four options that first involve migration, diversification, and storage.

Two further observations will highlight differences between Redding's approach and our own. First, Redding assumes that migration, diversification, storage and finally agriculture are interchangeable means of risk buffering as concerns their properties and that they differ only in their (increasing) costs. We would argue (see above and Halstead and O'Shea 1989) that each of these techniques is suited to different spatiotemporal patterns of resource fluctuations. There is no reason to expect a group in a particular environment to find all of them effective (if differentially costly) or to adopt them in the sequence given. Second, for Redding plant cultivation is the solution to subsistence risk. In our approach the shift to cultivation changes the ecological and economic nature of risk but does not eliminate it. Plant cultivation has its own risks and uncertainties: perhaps exceptional ones early in its evolution. We are skeptical that an activity with unpredictable success over a long production interval can be an effective solution to the short-term fluctuations of foraging success. We prefer to view risk as a condition of enduring causal importance but differing consequences throughout the transition from food gathering to food production.

The third model we mention is that of Layton et al. (1991; see follow-up comments in Hawkes and O'Connell 1992; Layton and Foley 1992; Winterhalder and Goland 1993). Layton et al. (1991) examine how different social and natural environments produce economies based on foraging, herding, and cultivation or some flexible combination of these subsistence practices. Their discussion brings together an eclectic mixture of environmental and sociopolitical elements that includes optimal foraging theory. They state, "If the cost of relying on high-ranking foods becomes too great, it should be worthwhile for foragers to add lower-ranking foods to their diet" (Layton et al. 1991:255). In their view, various consequences then may follow: the low-ranked resources may become subjected to cultivation in order to increase their yield, the distribution and predictability of the resource base may change such that territoriality becomes a more favorable option, or the human population may increase (Layton et al. 1991:257–60). The last hypothesis is a prediction that also occurs in our model.[11]

In a reply, Hawkes and O'Connell (1992) show that with closer attention to foraging theory one can question, sharpen, or extend on technical grounds several of the conclusions of Layton et al. (1991). Most of the points made by Hawkes and O'Connell are well taken, but we believe that they are mistaken (and that Layton et al. were correct in their original paper) on the issue of population. Hawkes and O'Connell (1992:64) say this: "Increases in diet breadth result from reduced foraging return rates and so lead to declines in population growth rates . . . sharper growth should be associated not with broader diets but with subsequent increases in handling efficiency associated with practices which result in domestication." We agree that declining foraging rates lead to increased diet breadth, but we believe that they also are consistent with enhanced forager population growth and increased density. The marginal reduction in foraging efficiency that brings such a resource into the diet is less important for population growth than the absolute magnitude of its sustainable yield and hence the population growth it will support (Winterhalder and Goland 1993).

Thus population growth may be associated with broader diets (even at lower foraging efficiencies) as Layton et al. (1991) predict. This effect will be augmented if there also are increases in handling efficiency as a consequence of coevolution and domestication. This observation allows us to refine a prediction of the diet breadth model that states that so long as highly ranked resources are sufficiently abundant, low-ranked items outside of the optimal diet will be ignored whatever their density. We add that they will also be ignored whatever their sustainable yield (which is a function both of density and intrinsic rate of increase, r). Thus they will be ignored whatever their potential consequences for forager population size. Put another way, from the perspective of foraging theory diet breadth and associated foraging efficiencies are intensive variables subject to (optimizing) selection. In contrast, the ultimate population-level consequences of incorporating a particular resource into the diet are extensive variables subject to the qualifications of Principle 3 (above).

Comparative Production Efficiencies of Foraging and Farming

We conclude with a comment on an issue that more than any other has framed and, we believe, has caused mischief in analyses of agricultural origins: the relative production efficiencies of foraging and farming.

Early models of domestication and agricultural origins assumed that food production was inherently more efficient than food gathering. Comparative evidence suggested that agriculture had such advantages of efficiency that it was an adaptation "waiting to happen," poised for the right combination of human preparedness and environmental circumstance. From this perspective one asked, Why did it take so long? However, with the dramatic revision of anthropological perceptions of hunter-gatherers in the early 1970s the comparative evidence was inverted. Foragers became the "original affluent society" (Sahlins 1972). Suddenly, the energy efficiency and limited effort of foraging looked quite good relative to the low productivity and high labor requirements of primitive agriculture. The new question became, Why would agriculture happen at all? Indeed, prime movers became critical to hypotheses on agricultural origins as anthropologists sought external variables capable of pushing socioeconomic development against the impedance of declining per capita energy efficiency. Why besides externally imposed necessity would foragers give up a secure and comfortable means of production for the uncertainty and drudgery of agriculture?

From an evolutionary ecology perspective either way of framing the question misleads the analysis. Whichever way the comparison falls, the juxtaposition of before and after production efficiencies to set the causal question places extensive variables at the heart of an evolutionary analysis. This violates Principle 3. The long temporal separation of the economies that define the before/after ratio makes it unlikely that it has causal evolutionary significance. No generation of foragers evolving to food producers faced choices in terms of the relative efficiencies attributed to the two distinct production systems. Energy production and efficiency may have mattered greatly, but they are properly set in terms of intensive variables (e.g., foraging efficiency), and these were generation-by-generation matters of selection among options with marginal differences.

Summary and Discussion

Our speculations to this point have been framed implicitly in the context of seasonal, temperate-zone agriculture, and with reference to "immediate return" (Woodburn 1982) hunter-gatherer societies. Other ecological situations (which differentially affect production interval, dispersion opportunity, and other variables identified here as important) presumably will produce variant outcomes. For instance, heavy use of traps or weirs allows a forager to disperse simultaneous production opportunities over space and thus to act more like a cultivator in adopting intrahousehold risk management. Sharing should be of less importance in such a society. Or, inasmuch as it is aseasonal, with the potential of a continuous harvest, tropical horticulture mimics the short duration of a forager's production interval. Despite heavy reliance on plant domesticates in the diet we might expect these horticultural systems to appear more like foraging societies than temperate-zone food producers. "Delayed return" (Woodburn 1982) hunter-gatherer societies that rely heavily on "seasonal" produce (e.g., salmon runs on the Pacific Northwest coast or acorns in central California) might have socioecologic structures more like those of temperate-zone farmers because of a long production interval. To the extent that evidence for ethnographic diversity can be related to this kind of ecological variability, it will help to test evolutionary ecology approaches to domestication and agricultural origins.

With respect to the processes that we propose accompany the prehistoric origins of agriculture, these summary points are worth emphasis:

1. The profitability (ranking) of a resource should be analyzed separately from its yield (density and intrinsic rate of increase) because these two variables have quite different ecological consequences. The differing evolutionary impacts of seed crops and maize in eastern North America may be an instance of this distinction.

2. The incorporation of low-rank but high-yield species into the diet forms the basis for population growth and creates a situation in which resources of higher rank will be heavily exploited and possibly depleted. Conversely, high-rank, high-density resources will exclude previously consumed items from the diet, allowing them to recover to preexploitation population levels.

3. Evolutionary ecology models show that local or regional pooling from spatially unsynchronized food sources is an effective means of reducing subsistence risk for both foragers and farmers.

4. Subsistence risk is likely to increase as high-yield domesticates enter the diet of foragers-becoming-farmers. An increasingly dense human population either overexploits alternative foods (if the TD is at low rank in the diet), thus decreasing effective dietary diversity, or (if the TD is at high rank) may grow to such an extent that previously foraged foods can provide only a modest buffer should the domesticate food or foods fail.

5. Growing dependence on plant cultivation changes both the subsistence production interval and dispersion opportunities in a manner that favors a shift from interhousehold (sharing) to intrahousehold (field dispersion) as a primary means of avoiding risk. Regional exchange remains a secondary risk-buffering mechanism.

6. Early agricultural households may be under strong selection pressure to develop community-level means of regulating rights in dispersed parcels of land. We speculate that they will develop mechanisms similar to those of the sectorial rotation systems seen in existing, nonindustrial agricultural systems.

7. We expect a lag in the development of risk-management strategies appropriate to increased reliance on high-yield, cultivated resources. The prehistoric record of increased paleopathologic conditions caused by nutritional insults during the transition to food production is consistent with this expectation.

8. The evolutionary ecology approach we advocate for explaining domestication focuses on selection-based models, nonnormative

Diet Choice, Risk, and Plant Domestication

properties of environmental variables, and the types of resource decisions actually faced by individual foragers. We propose that a microecological approach based on intensive variables is preferable to prime-mover analyses based on extensive variables, normatively described.

9. We have used literature summaries from the Great Plains and eastern North America to suggest how current archaeological evidence on processes of domestication and risk management might be interpreted using foraging theory. Although very preliminary, we hope that this material demonstrates the potential of an evolutionary ecology framework, for these and other regions of the world.

Acknowledgments

We thank Sheryl Gerety, Raymond Hames, Abram Kaplan, Bruce Smith, and Eric Smith for their helpful comments and criticisms of earlier drafts of this chapter. We are especially grateful to Richard Yarnell, colleague and teacher, for inspiration and gentle guidance in this and other endeavors.

Notes

1. By *normative* we refer to central tendency variables such as averages.

2. If environmental fluctuations are invoked, typically they are the regular ones of seasonality as they affect the scheduling of resource procurement (Flannery 1971) or transhumant migrations (Lynch 1973). Flannery and Reynolds (Flannery, ed. 1986) and Redding (1988) provide exceptions.

3. We use the term *selection* broadly and with some deliberate ambiguity. For this analysis is it unnecessary to choose between natural selection acting on genetic dispositions, cultural selection arising from adaptive biases in cultural inheritance, or the rational decisions of individuals (see Blurton Jones 1990; Durham 1992). Whatever the mechanism or combination, differential inheritance of the more advantageous option or options will produce cumulative, evolutionary change.

4. We refer here to the fine-grained, encounter-contingent model for prey selection, not the patch-choice model.

5. Although the diet breadth (defined as the number of resource types) will not narrow, the effective breadth (measured by both the number of types and the relative evenness of their representation in the diet) will drop.

6. This measurement is very close to McCloskey's estimate of 10%.

7. Three other families were at 100% probability of disaster for all possible numbers of fields. These are families with unusually high minimum requirements (because of factors such as shortage of potato land, large family size, or disastrously poor yields from other crops). No single field—or combination of fields—yielded well enough that it could have met their needs. Another three families actually experienced increasing probability of disaster with greater numbers of fields planted. These families provide an illustration of the second half of the rule of thumb: since their critical threshold is greater than the average yield, they might have done better to plant fewer fields (the high variance option). For these farmers, however, the choice is somewhat hypothetical. It would require that they know far in advance that their average yield would be less than their minimum needs. Lacking this prescience, these families bet as they had to on a reasonably good year, adopted risk-minimizing tactics accordingly, and then had the fate to experience increasing probability of disaster with increasing numbers of fields. The last family experienced 0% probability of disaster for all numbers of fields. This was the family with the smallest requirement, because of ample potato land and good yields among its other crops.

8. A denser stand of a grain crop may become more profitable because of increased efficiency in handling costs. This is quite unlike any benefit that would be gained, for example, in hunting, where the handling costs of a large ungulate will remain the same whether it is solitary or among a large herd at the time it is harvested. Following the diet breadth model, once the plant resource enters the diet, it will always be pursued once encountered. If such resources become physically associated with human settlements (by virtue of their establishment in disturbed habitats) then they will obviously be frequently encountered.

9. The decrease in health is not necessarily a direct result of changed dietary and nutritional circumstances. Diminished health may have resulted from population aggregation, which favors the spread of infectious diseases. Synergy between nutritional health and infectious disease may also have been important.

10. This citation is a mild instance of optimophobia, an affliction that causes one to be highly critical of optimality analysis even while practicing it in informal, implicit, or unrecognized forms (such as functionalist or adaptation approaches). It is an affliction more common among anthropologists than among economists.

11. The magnitude of the population increase will depend on the density of the resource.

CHAPTER 8

The Ecological Structure and Behavioral Implications of Mast Exploitation Strategies

PAUL S. GARDNER

The native nuts of the Eastern Woodlands have long been recognized to have been important foods of prehistoric Native Americans. Early descriptions of Native American diet in the Eastern Woodlands inevitably mention nuts. For example, during the De Soto entrada, "walnuts" (presumably thick-shelled hickories [*Carya* sp.] and pecan [*Carya illinoiensis*]) along with maize (*Zea mays* ssp. *mays*) and beans (*Phaseolus vulgaris*) were the foods mentioned most frequently by the Gentleman of Elvas as abundant in the native towns (Milanich 1991:70, 107, 131, 132, 136, 201, 202). Thomas Hariot's description of the sixteenth-century Carolina Algonquians includes accounts of their use of walnut (*Juglans nigra*), hickory, acorn (*Quercus* sp.), and chinquapin (*Castanea pumila*). Only acorns and chinquapins seem to have been used for bread (Quinn 1955:350, 354). All four types of nuts seem to have been boiled to make "spoonmeat" and nut "milk" (Quinn 1955:350, 351, 354), and all but acorn were eaten raw (Quinn 1955:350, 351). John Smith's description of nut usage among the early seventeenth-century Virginia Algonquians (Barbour 1986:151–53) largely reinforces Hariot's accounts but adds the information that the late spring–early summer was a prime period of nut consumption (Barbour 1986:162). Nuts remain prominent in late eighteenth-century accounts of southeastern Native American diet (e.g., Harper 1958:25; Williams 1930:439).

Furthermore, nutshell is very common in the Eastern Woodlands archaeological record, and archaeologists have long considered that gathering nuts was an important prehistoric subsistence activity (Caldwell 1958). However, the benefits derived from and problems entailed by exploiting nuts are rarely considered. In this chapter I examine some of the characteristics of nuts that seem important for structuring their exploitation. I conclude by drawing attention to some ways in which currently accepted models of Eastern Woodlands prehistory might be informed by a closer consideration of mast exploitation strategies. In this chapter I focus on hickory nuts and acorns, as they seem to have been the most important nuts, and data concerning them are most nearly complete.

Nutritional Composition

The nutritional composition of hickory nuts and acorns is well studied (United States Department of Agriculture [USDA] 1984). Hickory is the better source of calories. To meet the Food and Agricultural Organization's (1974) recommended daily individual energy intake of about 2200 kcal, about 12 ounces (340 g) dry weight of hickory nutmeat is required. About 15 ounces (427 g) of acorn is required. For comparison, about 1 1/3 pounds (604 g) of maize is required.

The reason for the relatively greater caloric content of hickory nutmeats is apparent when the proximate composition of the foods is considered (figure 8.1). Whereas maize and acorns are high in carbohydrates, hickory nutmeat has a much higher fat content. While obtaining an adequate supply of dietary fat is hardly a problem for affluent societies such as our own, there is reason to think that it may be for foraging peoples. Ethnographically, the Aché of Paraguay have been observed to collect particularly fatty foods even though doing so reduces their overall foraging efficiency (Hill 1988). In addition Speth and Spielmann (1983) have argued that prehistoric North American foragers whose diets were focused on ungulates such as deer may have experienced significant fat shortages in late winter and early spring because of the leanness of the game. The dietary fats provided by stored mast, hickory nuts in particular, may have been of considerable nutritional importance to Eastern Woodlands foragers.

For a plant food, hickories are a relatively good source of essential amino acids. The nuts are superior to both maize and acorns as a source of nine of the ten essential amino acids (figure 8.2). Acorn, however, provides even lesser amounts of seven essential amino acids than does maize, a notoriously poor protein source. Overall, hickory nuts would seem to be of more nutritional value to foragers than would acorns.

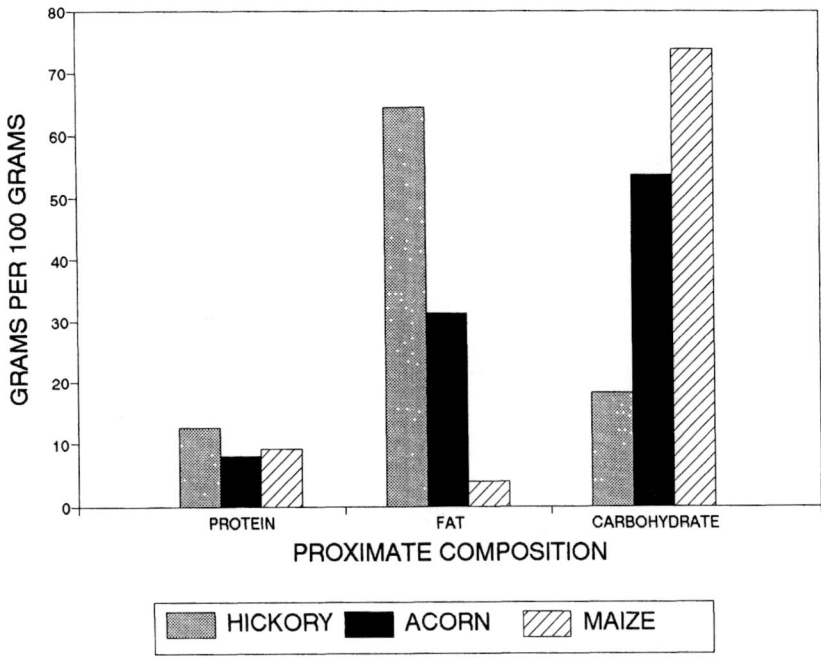

Figure 8.1. Proximate composition of hickory nuts, acorn, and maize (USDA 1984).

Yield and Availability

Potentially, hickories and oaks are enormously productive. Shell-bark hickory (*Carya laciniosa*) has been known to produce 3 bushels of nuts per tree (Schopmeyer 1974:271). Assuming 40 pounds (18 kg) of cleaned seed per bushel (Schopmeyer 1974:271) and a 35% edible portion (Watt and Merrill 1975), one can calculate a per-tree yield of 42 pounds (19 kg) of edible nutmeat. This amount of hickory nutmeat would provide 2200 kcal a day for fifty-six days. At this level of productivity, seven trees would feed a person for a year. Similar calculations can be made for acorn yields (USDA Forest Service 1948). Considering the abundance of oaks and hickories in most of the Eastern Woodlands (Braun 1950), it is easy to derive nutmeat yield figures that make the typical site catchment seem like Elysium (e.g., see Zawacki and Hausfater 1969).

Fortunately, more reasonable figures of nutmeat production are

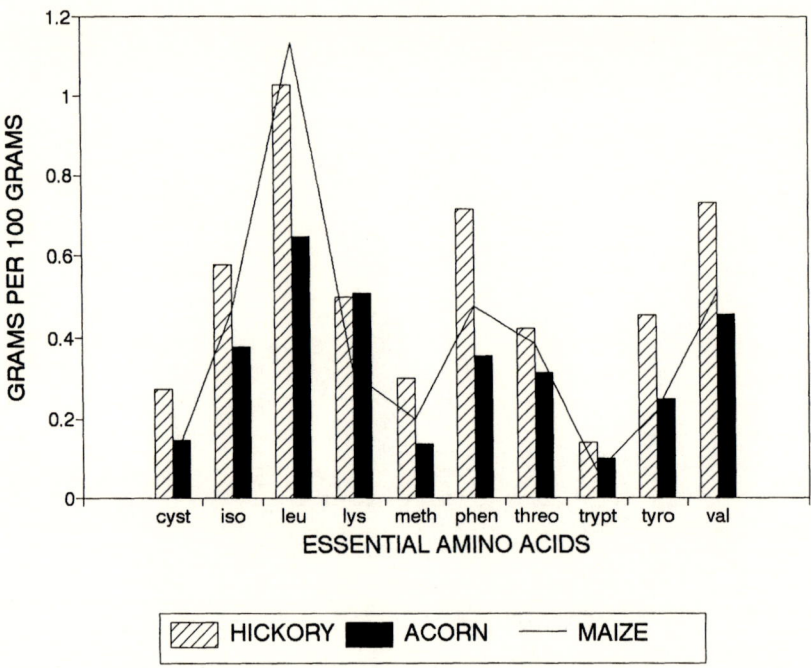

Figure 8.2. Essential amino acid composition of hickory nuts, acorn, and maize (USDA 1984). Key: cyst = cysteine, iso = isoleucine, leu = leucine, lys = lysine, meth = methionine, phen = phenylalanine, threo = threonine, trypt = tryptophan, tyro = tyrosine, val = valine (USDA 1984).

available from the wildlife management literature. Multiyear studies in Arkansas (Segelquist and Green 1968) and Ohio (Nixon et al. 1975) indicated an average acorn meat yield of about 53 pounds/acre (60 kg/ha). A multiyear study of hickory nut production in Ohio (Nixon et al. 1975) indicated an average nutmeat yield of 6.6 pounds/acre (7.4 kg/ha). However, extrapolating to a catchment of 10 km radius with even these carefully derived yields indicates an average acorn production of more than 4 million pounds (1.9 million kg) and a hickory nutmeat production of more than 500,000 pounds (236,000 kg). Obviously, the oak-hickory forest does produce, on average, a wealth of edible nuts. Thus exploitation of nuts by human foragers would not seem to be constrained by average forest productivity. Other factors limit the role of mast in human diets.

Constraints on Nut Exploitation

Processing Costs

Although nuts are nutritious and potentially abundant, all require some processing to render them edible. Minimally, all nuts require removal of the shells from the edible kernels. Acorns, in addition, must be leached to remove bitter tanic acids from the nutmeats. Unfortunately, little quantitative information is available from the ethnographic and historical records on the time costs of processing nuts. On the Hupa reservation in 1937, Goldschmidt (1974) observed that preparing acorn mush from 6 pounds (2.7 kg) of shelled acorns required two hours of pounding into meal, four hours of leaching, and an additional half hour of cooking. The end product was nearly 46 pounds (21 kg) of wet mush. Basgall (1987:29) using Goldschmidt's figures and estimating gathering, transport, and shelling costs calculated a total handling time for acorns of 4.2 hours/kg of nuts (1.9 hours/pound) for a caloric return of 1073 kcal/hour.

In a series of experimental replications of acorn collecting and shelling, Petruso and Wickens (1984) found that their time costs ranged from a low of 0.7 hour/kg for bur oak acorns to a high of 8.3 hours/kg for white oak acorns. The caloric returns were calculated as ranging from 290 kcal/hour up to 3500 kcal/hour (Petruso and Wickens 1984:367). These costs do not include any allowances for leaching or cooking. Reidhead (1976), combining Petruso and Wickens's data on collecting costs with his own processing experiments, calculated time costs of 1.6 hours/kg of processed bitter acorns. The caloric return was calculated as 2940 kcal/hour (Reidhead 1976:299). Reidhead's cost estimate included a 0.15 hour/kg cost for leaching pounded acorns in running streams (1979:235), a practice that is much more cost efficient than actively boiling the meal.

The costs of collecting and processing walnuts (*Juglans nigra*), hazelnuts (*Corylus* sp.), butternuts (*Juglans cinerea*), and hickory nuts (*Carya* sp.) have also been assessed experimentally. Talalay et al. (1984) processed walnuts, butternuts, and hazels by cracking the nuts and picking the nutmeats free, while hickory was processed by crushing and boiling the nuts as well as by the cracking-and-picking method. They determined that only hickories are suitable for processing by boiling, since the adherent husks of walnuts and butternuts contaminate the nutmeats during the boiling process, while hazel kernels will not float free.

The experiments of Talalay et al. (1984:356) indicated a kilogram of walnut kernels would require 10.5 hours of collecting and processing, a kilogram of hazel nutmeat would require 11.5 hours, and a kilogram of butternut kernels would require 28.6 hours. The caloric returns were 621 kcal/hour, 592 kcal/hour, and 247 kcal/hour, respectively. Collecting, cracking, and picking three species of hickory required a mean investment of 43 hours/kg of nutmeat. Collecting, crushing with a nutting stone, and boiling the three species of hickory required an average of 3.5 hours/kg. Relying on experiments performed by Reidhead (1976), Talalay et al. calculated the mean cost of collecting, crushing with a mortar and pestle, and boiling as 2.0 hours/kg of nutmeat. The mean caloric returns using the three techniques of processing hickory were 165 kcal/hour, 2030 kcal/hour, and 3480 kcal/hour, respectively (Talalay et al. 1984:356). They concluded that butternut exploitation was probably economically impractical because of its low caloric returns and that walnuts and hazelnuts were probably of only limited caloric benefit. In addition, they concluded that hickory nuts would require a crushing and boiling technology to make them calorically worthwhile (Talalay et al. 1984:356).

Experimental studies such as the above are very useful for the insights they provide into prehistoric subsistence systems and are invaluable for the development of quantitative models. However, a number of factors dictate that they be used with caution.

1. It is doubtful that experimenters undertaking novel tasks can replicate the activities with the same skill and efficiency as experienced foragers. In addition, there can be little assurance that the range of subsistence tasks replicated experimentally exhausts the possible means of exploiting a resource. For example, the dichotomy of cracking and picking versus crushing and boiling the hickory nuts omits the historically described practice of separating finely pulverized hickory shell from the nutmeats by eating it all and spitting out the shell (Lefler 1967:105). Not only would this seem more energetically efficient than either of the other two practices, it would also explain the frequent presence of hickory nutshell in human paleofeces (Yarnell 1969).

2. Generally, the collecting experiments are conducted over brief

time intervals, often as short as five minutes (Petruso and Wickens 1984:366). It is not at all clear that the yields measured are in fact sustainable over long periods of collecting. In addition, given the great variation in nut yields from year to year and from tree to tree (Downs and McQuilkin 1944; Nixon et al. 1975; Sork 1983), an "average" return may have only limited relevance to decisions made by highly knowledgeable and presumably highly selective human foragers.

3. Given the great transformation of the Eastern Woodlands following European contact, there seems to be no way of assessing how closely modern collecting sites resemble the aboriginal ones. Collecting from modern trees in parklike habitats may artificially inflate yields. Collecting from modern trees in cutover, second-growth forests may artificially deflate yields. In addition, there is no way of assessing to what extent Native Americans managed the forest to create productive tree stands (Munson 1986).

4. The experiments tend to focus on key aspects of foraging such as collecting and processing but omit other aspects such as travel costs. If tree stands produce small yields relative to human needs, traveling between stands would be a critical component of foraging costs. In addition, if stands are far removed from the residential bases, transporting the gathered nuts back to the storage areas might well be the most arduous and time-consuming task associated with collecting. Hence the distribution of the productive trees on the landscape may have been of critical importance for determining foraging costs.

Again, although ethnoarchaeological studies of Native American subsistence techniques are valuable, models of prehistoric subsistence based on them must be evaluated critically.

Periodicity of Yields

Although mast is nutritious and sometimes abundant, some reports suggest that it has a limited value since nut trees bear at irregular intervals. It is true that it is rare to get two good harvests in a row. Some of the variation in mast yields is a result of the effects of weather. Cold weather in the spring during the time of pollination and fruit set can result in poor yields (Goodrum et al. 1971). The effects on white oak of a harsh spring in 1966 in southeastern Ohio can be seen in figure 8.3.

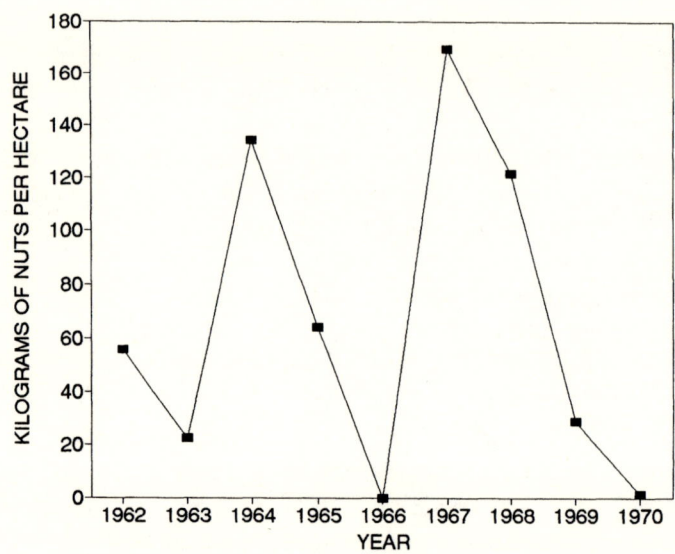

Figure 8.3. Annual yields of white oaks in southeastern Ohio (Nixon et al. 1975).

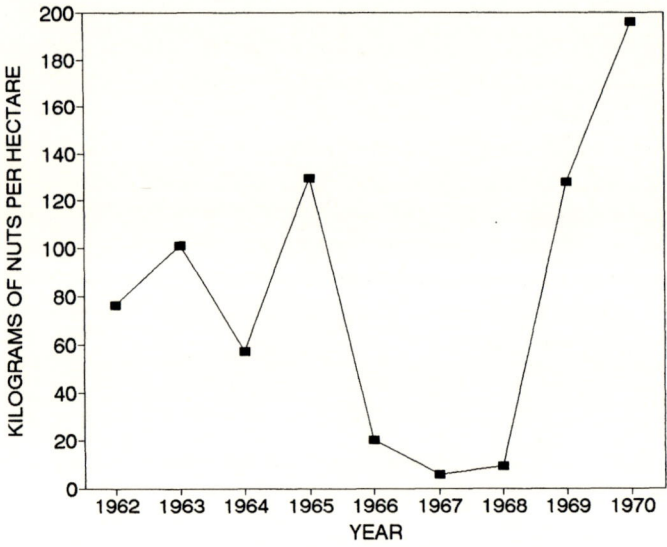

Figure 8.4. Annual yields of red oaks in southeastern Ohio (Nixon et al. 1975).

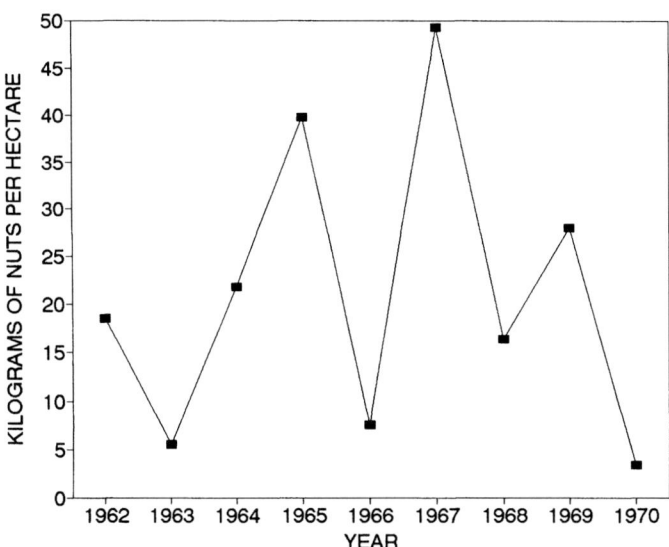

Figure 8.5. Annual yields of hickory nuts in southeastern Ohio (Nixon et al. 1975).

In red oak the additional effects were seen the next year, since red oak acorns take two years to develop (figure 8.4).

In addition to the variations produced by weather, some of the variation in annual mast yields is under genetic control. Presumably the frequent periodic low yields (figures 8.3 through 8.5) evolved as a defense tactic against seed predators (Sork 1983). Because of the genetic component, the fluctuations in yield are to some extent predictable.

A means of measuring the predictability of periodic phenomena using the Shannon information statistic has been suggested by the ecologist Robert K. Colwell. Colwell (1974) points out that predictability can be expressed as the sum of two components. The first is constancy, which is the tendency of a phenomenon to remain in the same state. In the case of mast production, constancy is the evenness of the annual production. The second component of predictability is contingency, which is the degree to which particular states are correlated with particular periods of the cycles. In the case of mast production, this reflects the extent to which abundant crops occur at regular intervals. Figure 8.6 shows a plot of constancy and contingency of mast yields for two-year cycles in the southeastern Ohio study area over a nine-year period (data from Nixon et al. 1975).[1] In the graph, predict-

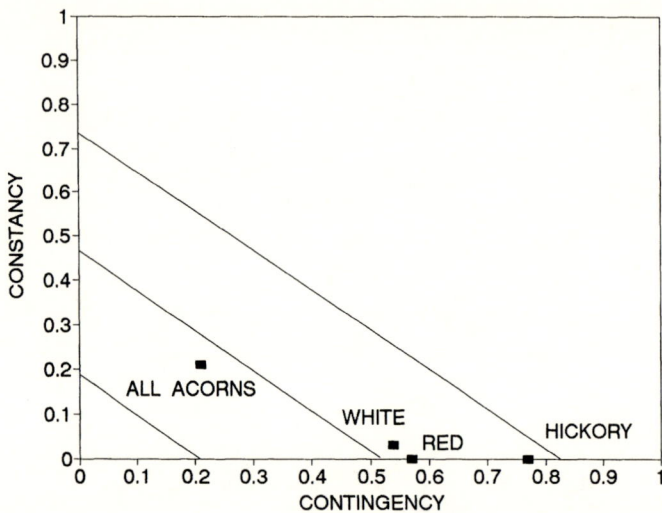

Figure 8.6. Constancy and contingency scores of acorn and hickory nut yields in southeastern Ohio.

ability increases away from the origin, and the diagonal lines connect points of equal predictability. Constancy is very low for both red and white oaks and for hickories. In fact, for hickories and red oaks the constancy score is a dead zero, since nut yields fluctuate so markedly. Constancy does rise somewhat when acorns of both red and white oaks are considered together. This is congruent with anecdotal reports that good crops in one type of oak often buffer poor crops in the other.

The figures for contingency show hickory nut yields to have been the most regularly periodic. As a result, hickory has the highest overall predictability. The combined acorn contingency is the lowest of all. This result may reflect the fact that the genetically induced variation in yields of the two types of oaks have largely canceled each other out, leaving only the more stochastic variation caused by weather. Obviously these predictability coefficients are hardly definitive, but these data do suggest that the problem with relying on mast as a food is not so much unpredictability per se, but rather that shortfalls in yields are so frequent.

Harvest Period

As mentioned above, the eastern forests are capable of yielding enormous amounts of mast. However, calculations of average forest yields probably have little relevance to investigations of human subsistence. First, all published yield figures are probably best considered maximums, since numerous factors such as animal competition, insect infestation, mold, and rot combine to reduce the number of edible nuts available to human beings (Lopinot 1982:724). Second, the availability of nuts is probably limited not so much by the number produced, but rather by the amount of time during which they may be gathered (Asch and Asch 1979). Usually mast disappears from the forest floor within a few weeks after it falls because of exploitation by woodland animals (Downs and McQuilkin 1944; Petruso and Wickens 1984; Talalay et al. 1984). Hence, during the limited period of nut-fall, it seems most likely that human foragers would focus their foraging efforts on only the most productive nut stands while ignoring the "average" trees. Since there is considerable variation among the yields of individual trees, and since the variation in relative yield is consistent from year to year (Downs and McQuilkin 1944; Sork 1983), Native Americans should have been able to plan selective exploitation with considerable effectiveness. In addition, since per-tree yields are correlated with crown development and exposure to sunlight (Downs and McQuilkin 1944; Goodrum et al. 1971), there would seem to be great potential for human foragers to inflate per-tree yields by thinning the forest around productive trees (Munson 1986). By managing tree stands in this fashion, human foragers would not only increase per-tree yields, but would also eliminate the need to search the forest each fall for productive trees. This would facilitate scheduling of mast collection and allow the limited harvest season of mast to be utilized most effectively.

Seasonality and Storage

Nuts, of course, become available in the fall. However, fall would be the time of relative plenty for Eastern Woodlands foragers: game is generally fat and abundant, deer are rutting and especially vulnerable, and many native fruits are ripe. Thus, in the fall, mast may have been just one resource among many. In the late winter and early spring, however, other resources would have been harder to come by. Game is lean and fewer in number because of winter mortality, deer become increasingly solitary and less mobile as the fawning season approaches, and fewer plants are available. I suggest that mast may have been of more importance as a stored commodity in the winter and spring than as a fresh food in the fall. Some support for this suggestion comes from John Smith's observation that mast was of particular importance to Virginia Algonquians in May and June (Barbour 1986:162).

However, as mentioned above, mast is such an important food to so many species, it usually disappears from the forest floor within a few weeks after it falls (Downs and McQuilkin 1944; Petruso and Wickens 1984; Talalay et al. 1984). Hence if mast is to play more than a short seasonal role in the diet, then storage would seem necessary. Fortunately, nuts are eminently storable resources: they come prepackaged in their own individual containers. No elaborate preparation is required to store nuts other than parching them, which precludes sprouting and kills any molds, fungi, or insects that infest them.

Nuts are somewhat bulky, however, and it might be thought that this would make storage less feasible. Actually this is not the case. The high caloric content of nutmeat allows a significant number of calories to be stored in spite of the bulk produced by the inedible nutshell. For example, a pit 1 m deep by 1 m in diameter could hold about 22 bushels of hickory nuts. At 40 pounds/bushel (Schopmeyer 1974:271) and an edible portion of 35% (Watt and Merrill 1975), this represents more than 300 pounds of edible nutmeat. This amount of hickory would provide 2200 kcal daily for more than thirteen months. The same pit, if filled with red oak acorns, would contain about 680 pounds of acorn nutmeat, using a rough estimate of 50 pounds of acorns per bushel (Schopmeyer 1974:699) and a 62% edible portion (USDA 1984). This amount of acorn meat would provide 2200 kcal daily for nearly twenty-four months. If filled with white oak acorns, the pit would contain about 955 pounds of edible nutmeat, based on a rough estimate of 70 pounds of acorns per bushel (Schopmeyer 1974:699) and a 62% edible portion (USDA 1984). This amount of acorn meat would provide 2200 kcal daily for nearly thirty-three months. Overall, although mast is bulky, providing adequate storage facilities would not be a serious problem.

Some Implications of Mast Exploitation

To summarize to this point, I would argue that hickory nuts and acorns should have been desirable resources to all Eastern Woodlands foragers. Nutritionally, they complement lean game quite well. Although their yields are affected by poor weather, so are all other foraged foods. Mast yields are somewhat periodic, but not completely random. In addition, it would seem feasible to store a significant amount of mast—at least enough to ease any late winter–early spring hungry season. However, given the competition from other animals for mast, it seems unlikely that a strategy of residential mobility would allow mast to play anything other than a brief seasonal role in the diet.

Perhaps surprisingly, given the prominent role of nutshell in Archaic period sites (Asch and Asch 1979; Yarnell and Black 1985), the behavioral implications of exploiting mast are rarely considered in models of Archaic period subsistence-settlement systems. More often, such models attempt to relate broad environmental parameters with general mobility patterns (e.g., Binford 1980; Claggett and Cable 1982). The test implications of the proposed models typically focus on site size and structure, feature and artifact densities, and ratios of formal to expedient lithic tools (Anderson and Hanson 1988; Anderson and Schuldenrein 1983; Claggett and Cable 1982; Stafford 1985). Although "tools" figure prominently in these scenarios, little attention is paid to the organization of work in general or to the behaviors that would be necessary to exploit food resources in particular. I would suggest that our models of Archaic period subsistence might be usefully informed by a consideration of the behavioral constraints imposed on foragers by the ecological structure of mast resources.

Recently it has been suggested that production of hickory nut oil became integrated into a logistical collecting strategy in which hickories were collected and processed into oil at extractive camps before being transported to residential bases. Stafford (1991) identifies a number of midwestern sites with pit features and high nutshell/wood ratios as extractive camps used for hickory nut oil production. I agree that logistical collecting of hickory nuts seems most reasonable. I likewise think it probable that the sites identified by Stafford are nut extractive camps, but I do not think oil production was involved. First of all, whole nuts, although they are bulky, are still easier to transport and store than a liquid. For example, the same amount of hickory nuts that can be stored in a 1 m by 1 m pit yields about 11 gallons of oil (Battle 1922). While containing this might not be a problem for a sedentary, pottery-using society, I think it would be rather incommodious for Archaic period foragers. In addition, it seems likely that nut oil would be more prone to spoilage during the storage period than would be parched whole nuts. Finally, producing the nut oil would be time consuming.

Given the limited harvest season for nuts, about four weeks (Talalay et al. 1984), I think nuts are likely to have been collected and stored in the most expedient manner possible. This would involve collecting the nuts, then parching them to stabilize them for storage. Parching is a necessary step that kills any molds, fungi, or insects that inhabit the nuts and likewise kills the embryos of the nuts thereby precluding germination during storage (Talalay et al. 1984). Parching nuts could be readily accomplished by stirring the nuts over heated stones in shallow pits. This would leave features very similar to those characterized as nut boiling pits. In addition, I think the high density of nutshell at the sites probably results from the collectors gorging on the abundant nuts during the harvest period. In fact, it appears that the archaeological signature of a nut parching station would be basically identical to that of a nut boiling station.

Overall, a more detailed consideration of how the work of exploiting nuts must have been organized on the ground suggests that the addition of an added processing step during the critical harvesting season is unlikely to have been a routine practice. Likewise converting self-contained prepackaged nutmeats into a greasy, spoilage-prone liquid prior to transport and storage seems equally contrary to expected behavior. While production of nut oil at extractive camps is easily fitted into general models of logistical collecting, a consideration of "on the ground" activities makes it seem less practical.

In an intriguing paper, Munson (1986) has suggested that Middle Archaic period population aggregation, increased residential permanence, and increased hickory nut use all result from a technological change in hickory nut processing. In particular, Munson argues that the adoption of stone boiling as a technique for removing hickory nutmeats from the shells transformed hickories from one of the most costly to the least costly nut to process (he does not suggest that boiling was carried to the point of producing oil). In this scenario, the addition of abundant hickory nuts to the diet raised the carrying capacity of the local forests and allowed population aggregation and increased residential permanence in the prime river bottom habitats.

Although it is refreshing to see archaeobotanical data play a seminal role in an explanation of prehistoric culture change, several objections can be raised to Munson's hypothesis. First, were hickory so inefficient to utilize prior to stone boiling, one would expect it to be rare in earlier sites, when in fact hickory nutshell is very common in Early Archaic sites (Asch and Asch 1979, 1985a; Chapman and Shea 1981). Second, if stone boiling did have such dramatic adaptive consequences, one would expect the innovation to have spread rapidly across the Eastern Woodlands. In fact, Munson's graph of the relative contribution of hickory nut at Modoc Rockshelter indicates a sharp increase ca. 8300 B.P. (Munson 1986: fig. 2), whereas hickory does not sharply increase at Koster until about 7500 B.P. (Asch and Asch 1979). As the two sites are only about 150 km (94 miles) apart, a very slow rate of diffusion is indicated. Third, a technological advance in nut processing would seem likely to increase their consumption throughout the Eastern Woodlands. However, there is considerable regional variability in the prevalence of nutshell in Middle Archaic sites. For example, in the Little Tennessee River area, Middle Archaic period sites do not display an increased prevalence of nutshell; rather nuts seem to have become less important overall (Chapman and Shea 1981). Furthermore, the average contribution of hickory nutshell relative to other types in Middle Archaic sites increases to only 87% from 83% in Early Archaic sites (Chapman and Shea 1981). Certainly this pattern does not suggest a dramatic innovation in hickory processing. Finally, as mentioned previously, Munson's dichotomy of cracking and picking versus crushing and boiling does not exhaust the possible means of processing hickories. The "snarfing-and-spitting" technique described by Lawson (Lefler 1967: 105) is probably more efficient than either. Overall the "florescence" of hickory nut exploitation in the Middle Archaic seems unlikely to have resulted from a technological innovation.

Another, perhaps more parsimonious, explanation for increased hickory nut use in the Middle Archaic is climatic change, particularly the onset of the warmer and drier conditions of the Hypsithermal interval. Most hickories and oaks favor warm, dry conditions (Fowells 1965). During the Hypsithermal the absolute abundance of hickories and oaks may have increased in many locales, and perhaps more important, the warmer, drier conditions are likely to have made weather-related mast failures less common. With an increase in the overall abundance of nuts in the environment, foragers would have been able to specialize on the most favored nut, which I argue was hickory because of its nutritional superiority. Thus climatic change could account for the increased use of hickory nuts in the Middle Archaic.

In addition, there is evidence that the Hypsithermal might have been a more indirect cause of increased hickory exploitation at Middle

Archaic sites such as Koster. Geomorphological studies indicate that Hypsithermal-induced erosion in the uplands led to floodplain aggradation in the Illinois River valley (Hajic 1981), which produced resource-rich slack-water riverine habitats and floodplain lakes (Brown and Vierra 1983; Styles 1986). Middle Archaic foragers responded to this increased resource abundance by reducing residential mobility and establishing large, relatively sedentary occupations in the prime floodplain habitats (Brown and Vierra 1983). It seems reasonable to assume that these large, localized populations would have eventually experienced some dietary stress, perhaps particularly in the cold seasons when aquatic resources would have been less accessible. The abundant, nutritious, and easily stored hickory nut of the surrounding forests would seem an ideal candidate for augmenting the Middle Archaic diets. In this scenario, the exploitation of hickory nuts intensified, not because technological change made them easier to utilize, but because demographic change necessitated accepting relatively high production costs in order to acquire large amounts of a nutritious food.

It is interesting to speculate what would have transpired at the end of the Hypsithermal when the warm, dry conditions ended. I suggest that one of the first ways in which the return of cooler, moister conditions might have affected Eastern Woodlands foragers would have been an increase in the frequency of mast failures caused by spring frosts. Since weather-related failures of mast are likely to have occurred on a regional scale, one means of buffering them would have been through the establishment of interregional alliance networks. Interestingly, it is at the end of the Hypsithermal and the beginning of the Late Archaic period that we see evidence of appreciable interregional exchange in the Eastern Woodlands. These exchange networks may have provided important extralocal alliances for Eastern Woodlands foragers trying to exploit an increasingly unpredictable forest.

Additionally, since oaks and hickories prefer more xeric habitats, increasingly mesic conditions are likely to have forced them away from the prime river bottom locales. As the oak-hickory stands retreated upslope, it would have become more time consuming to gather nuts from base camps located in the major river bottoms. One means of dealing with this problem would be to disperse the population. Again, the general pattern of the archaeological record indicates an increased occupation of uplands and smaller tributary valleys during the Late Archaic period (Munson 1986).

Without the aquatic resources of the major river valleys, foragers in the less-productive uplands and smaller tributary valleys may have increasingly managed local mast-producing tree stands in order to promote greater yields. Oaks and hickories produce their biggest crops when they have large expansive canopies well exposed to sunlight (Downs and McQuilkin 1944). A simple way of creating this condition would be to open the forest canopy by ring girdling nonproductive trees. As the nonproductive trees died and dropped their leaves, productive trees would have received more sun and become larger and more productive (Munson 1986).

Opening the forest canopy to promote nut production would have had the unexpected consequence of increasing the available habitats for sun-loving weeds. Significantly, the creation of weedy anthropogenic habitats near human settlements has been specified as an early stage in the coevolution of plant-human relationships that leads to plant domestication (Rindos 1984). Bruce Smith (1987b) has posited that this stage was reached in the Eastern Woodlands in the Middle Archaic, when weedy anthropogenic habitats developed as a consequence of increased sedentism in prime riverine habitats. Although the prime riverine habitats undoubtedly supported the largest human populations and, hence, would have received the most unintentional anthropogenic disturbance, weedy habitats may have been created quite widely across the landscape as a consequence of the dispersal of foraging populations engaged in nut tree management. Furthermore, an expansion of the diet to include alternative foods such as weed seeds would seem as likely to occur in less favorable habitats supporting smaller populations as in prime habitats supporting large ones. Overall, I would argue that sites outside of the most productive river valleys are equally likely locations for the initial stages of plant husbandry. This may help explain why upland rockshelters produce some of our earliest evidence of cultigens (Cowan et al. 1981; Fritz, this volume; Gremillion 1993c and this volume; Smith and Cowan 1987).

In conclusion, I can point out that even if I am not correct, and Eastern Woodlands agriculture did not have its origins in an attempt to maintain mast productivities, still it did originate from the preceding foraging adaptation. Since the exploitation of mast is likely to have been an important component of that adaptation, more attention should be paid to understanding how mast exploitation articulated with other aspects of prehistoric culture. While tabulating nutshell percentages is not as flashy an endeavor as finding early cultigens, nutshell is an important data set in its own right. It is worthy of more from ethnobotanists than relegation to the denominator of seed/nutshell ratios.

Acknowledgments

This chapter has benefited from the comments and suggestions of Kris Gremillion, Lee Newsom, and an anonymous reviewer. A special thanks is owed to Dick Yarnell for encouraging my interest in cultural ecology, introducing me to archaeobotany, and otherwise saving this archaeologist from a life of ceramic analysis.

Note

1. The statistics were calculated by standarding the yield data from Nixon et al. 1975. Each annual yield was then scored as low ($z < -0.43$), average ($-0.43 < z < 0.43$), or high ($z > 0.43$). The constancy, contingency, and predictability of these three ordinal "states" were then calculated using the formula from Colwell (1974:1149–50).

CHAPTER 9

Changing Strategies of Indian Field Location in the Early Historic Southeast

GREGORY A. WASELKOV

During the historic period, from the mid-sixteenth century to the early nineteenth century, Indian societies of southeastern North America adjusted to changing epidemiological, demographic, political, technological, and social circumstances that developed in the course of European conquest and colonization. Native American adaptive responses to this invasion involved alterations in traditional agricultural practices, including the strategies employed in selecting field locations.

Direct evidence of Indian fields has only occasionally been retrievable from the archaeological record (Fowler 1969, 1992; Kelly 1938; Riley 1987), although recent conceptual breakthroughs in settlement interpretation promise better results in the future (Killion 1992). On the other hand, historical documents, both written descriptions and cartographic depictions, can provide considerable insight on preferred agricultural soils. This data source has not been exploited in a systematic way by southeastern ethnohistorians, but it could serve as a useful corrective to the most naive of cultural resource models that rely exclusively on site-soil correlations.

Because native agricultural goals and methods varied greatly across the Southeast, both among societies and through time, it is not yet possible, unfortunately, to account for all strategies of Indian field location according to any single, unified explanatory model. The western Muskogean Choctaws and Chickasaws seem to have followed a sequence of farming strategies that was out of synchrony with that of their congeners to the east. Historical analysis of Choctaw agricultural practices suggests that their late prehistoric preference for bottomland soils in major river valleys gave way, in the early historic period, to a decided partiality for upland loamy soils, only to revert in the nineteenth century to the bottomlands (White 1983:14, 24–25, 132–37).[1] The Chickasaws may have undergone a similar transition (Johnson et al. 1989; Waselkov 1989:332–34) (figure 9.1). Archaeologists are just beginning to consider the question of changing Choctaw and Chickasaw settlement systems, and we should be learning much more in the near future about their strategies of field location.

The focus of this discussion will be restricted to another part of the Southeast, from the southern Appalachian Cherokee country, through the Creek (or Muskogee) territory of modern-day Alabama, Georgia, and Florida, to the north-central Gulf coast. There, a different sequence of shifting field location strategies evidently occurred during the historic era. Throughout this portion of the Southeast, the seventeenth-century and eighteenth-century native inhabitants of the region seem to have continued the traditional agricultural practices of their Mississippian predecessors, employing a labor-intensive, mixed-habitat strategy that combined large, communal field cultivation in floodplain bottomlands with household gardening on terrace soils in or adjacent to villages (Hudson 1976:291; Ward 1965:44; Woods 1987). Since bottomlands are subject to periodic flooding, which is accompanied by silt deposition that maintains soil fertility indefinitely, fields located there were continuously cultivated for centuries.

When the French arrived on the north-central Gulf coast in 1699 to establish the colony of Louisiana, they found several small tribes (including Mobilians, Tomehs, and Pensacolas) living in villages situated in and around the Mobile-Tensaw delta, which at 180,000 acres is the third-largest swamp in the Southeast. Most of these groups, and other small refugee tribes that soon joined them, typically located their villages on high bluffs overlooking the delta and placed their fields in the bottomlands, which were inundated every spring (figures 9.2 and 9.3). At first, the French and Indian settlements coexisted as complementary segments of a single colonial system, with Indians providing most of the food for the French in exchange for manufactured goods imported from Europe and other European colonies in the Americas (Rowland and Sanders 1932:152; Zitomersky 1992:163–64). Later, French (and eventually British, Spanish, and American) settlers displaced the Indians by establishing plantations of their own. They accomplished this principally by appropriating bluff-top village sites for their farms and towns,

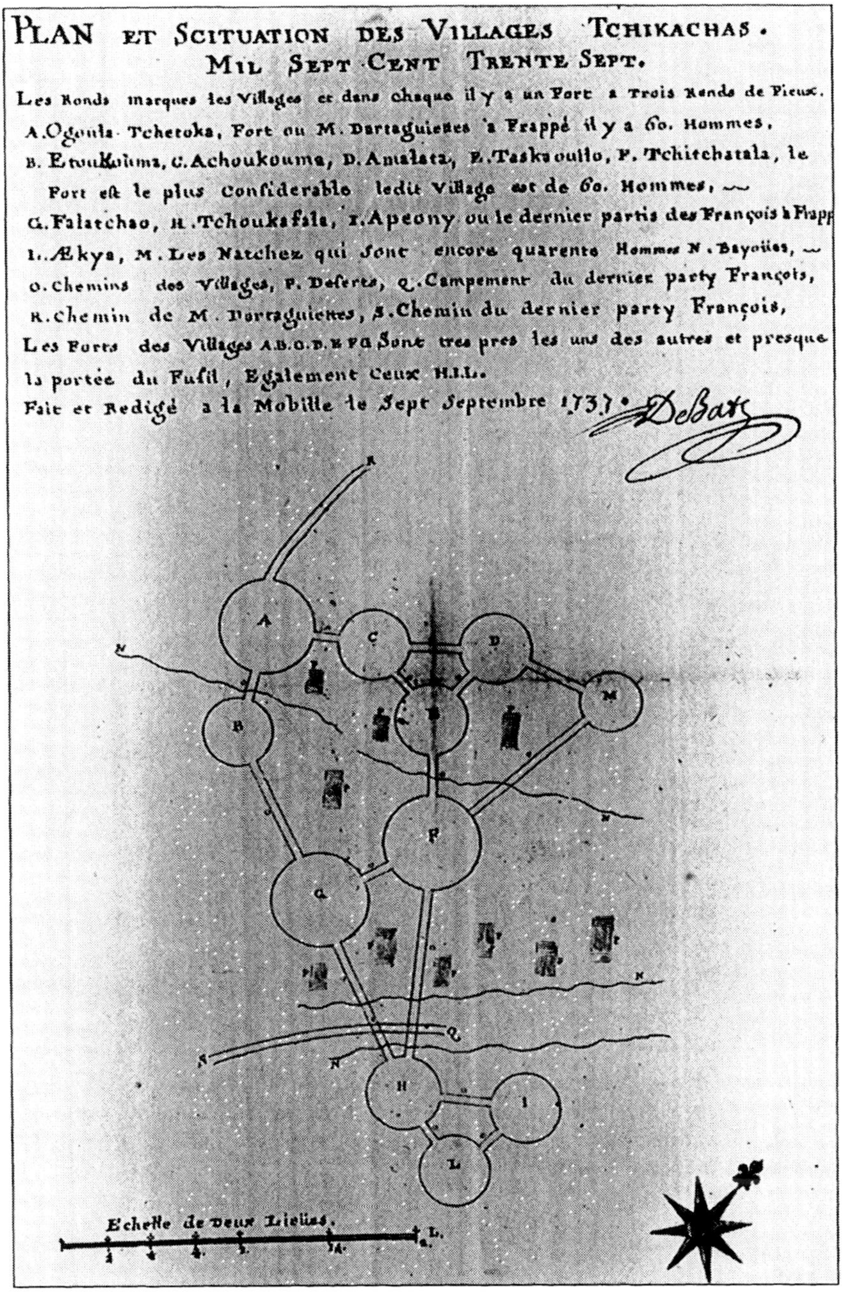

Figure 9.1. Alexandre de Batz's copy of an Alabama Indian map of the
Chickasaw villages, 1737. Chickasaw agricultural fields are the shaded rectangles
("Plan et scituation des villages Tchikatchas, par Alexandre De Batz." Plume et
encre, aquarelle sur papier, 7 septembre 1737. Archives nationales [France],
Centre des Archives d'Outre-mer, Aix-en-Provence. COL. C 13 A, vol. 22, fol.
68. Tous drois de reproduction réservés).

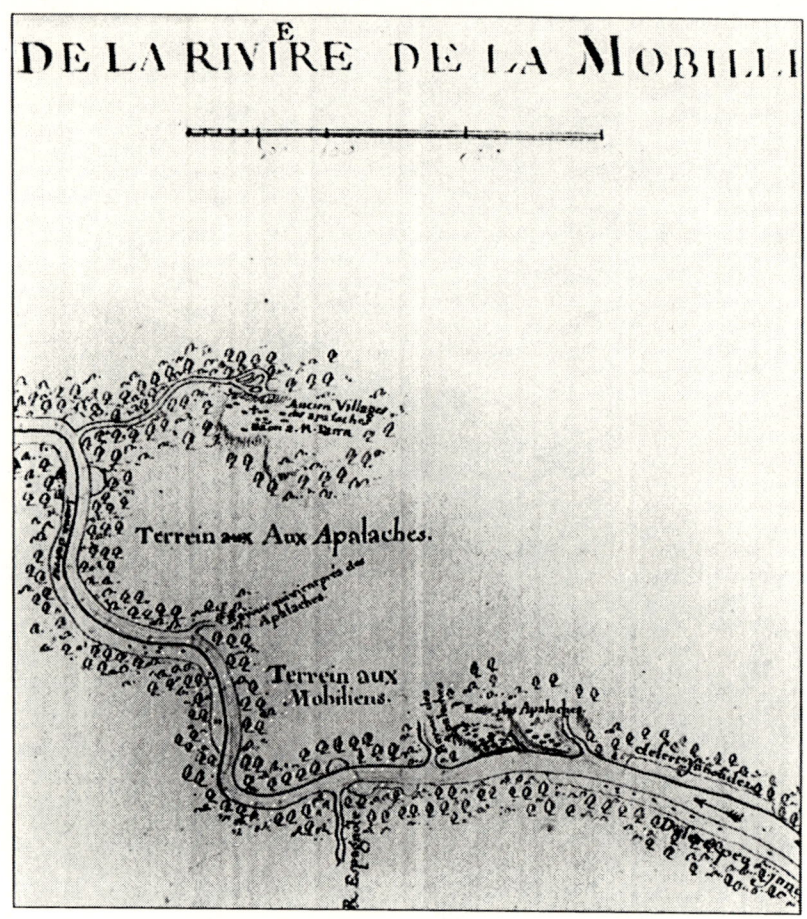

Figure 9.2. Detail of "Carte d'une partie du cours de la Mobille et de celle des Chicachas," circa 1725, showing Apalachee and Mobilian Indian fields as *desert yanondez*, inundated fields (Library of Congress, Geography and Map Division, Alabama, G3972.M6 1788?.C2 Vault).

as happened when the French relocated a Chatot village in 1711 to make way for their colonial capital, Mobile (Potter and Waselkov 1994:31). The Indians responded by moving their villages away from the familiar delta soils to the banks of small tributary streams, where they cleared new fields on the stream terraces and adjacent uplands. The bottomland old fields were eventually incorporated into colonial plantations once slave labor was adopted in Louisiana (Lankford 1983). By

Indian Field Location in the Early Southeast

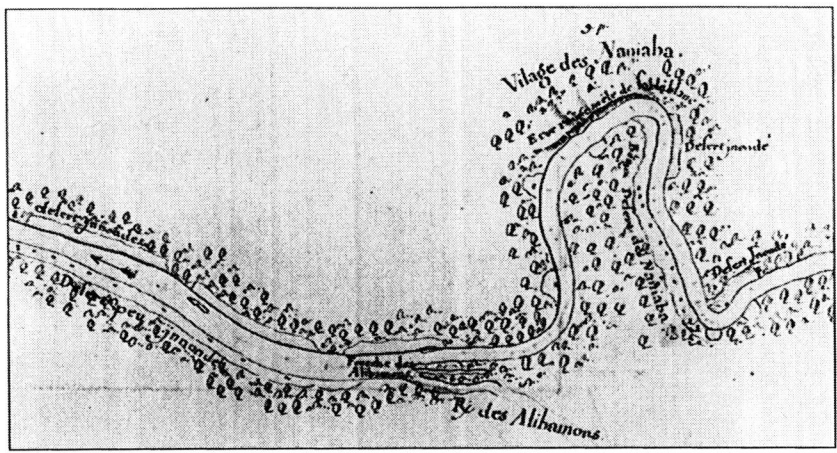

Figure 9.3. Detail of "Carte d'une partie du cours de la riviere de la Mobille et de celle des Chicachas," circa 1725, showing Naniaba Indian fields labeled as *desert jnondé,* inundated fields (Library of Congress, Geography and Map Division, Alabama, G3972.M6 1788?.C2 Vault).

1771, according to a map drawn in that year by David Taitt, only two Indian old fields remained in the delta, which had been subdivided into numerous European plantations (figure 9.4).

The usurpation of Indian fields in river bottomlands occurred repeatedly throughout the Southeast during the historic period as Indian villagers responded to direct demands for land from their increasingly numerous white neighbors. More complex situations developed in which populous groups such as the Creeks and Cherokees were able to limit (to some extent at least) the rate of white encroachment on their lands. Their territories encompassed a wide variety of habitats, but many of the largest Creek and Cherokee towns were located in major river valleys, where expansive bottomlands could accommodate extensive agricultural fields. As was the case with other early historic southeastern groups, Cherokee and Creek matrilineages controlled agricultural production in large, communally worked lineage fields and in household gardens located within the towns. While the labor of male hunters had been co-opted by the English deerskin trade as early as the late seventeenth century, Creek and Cherokee women successfully resisted the integration of their agricultural production into the Euro-American market economy until the early nineteenth century (Hatley 1989, 1993:8–9, 160–63, 233–34). This very important event (or, actually, a

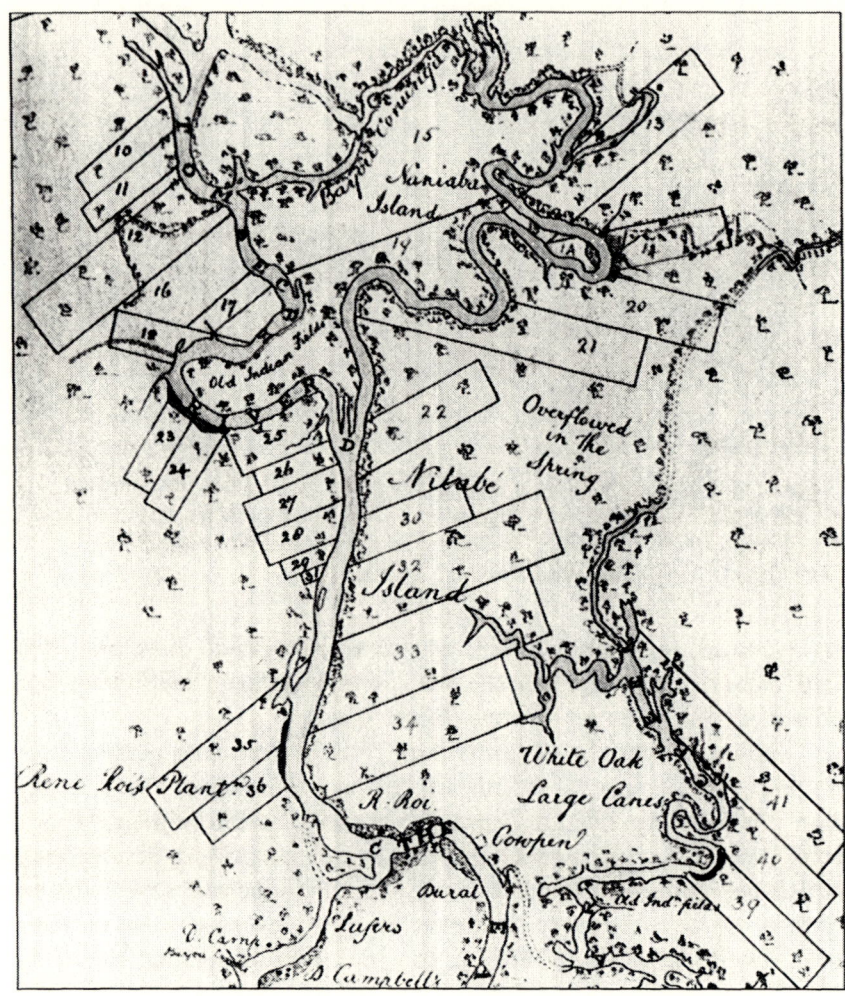

Figure 9.4. "Old Indian Fields" surrounded by English land grants in the Mobile-Tensaw delta area, as shown on "A Plan of Part of the Rivers Tombecbe, Alabama, Tensa, Perdido, & Scambia in the Province of West Florida," by David Taitt, 1771 (Library of Congress, Geography and Map Division, Alabama, G3971.P53 1771.T3 Vault).

process that occurred over several decades) is the key to understanding the concurrent change in the Creek and Cherokee strategy of field location.

For more than a century, hunters had obtained European-made goods in exchange for deerskins. Deer were hunted from community and tribal territories in quantities determined by the needs and abilities of individual hunters and their families. The manufactured goods they received for deerskins permitted them to substitute iron hoes and axes for stone counterparts, glass beads for shell, and cloth for leather. In fact, most of the newly acquired objects replaced prehistoric forms; few were truly innovative in function. Cherokee and Creek cultures selectively incorporated elements of European-made technology throughout the eighteenth century, while remaining distinctively Cherokee and Creek.

Beginning about the year 1790, the market value of deerskins declined in the United States and Europe (Coker and Watson 1986; Phillips 1961:vol. 2, 468–500). A few village headmen had begun keeping livestock as early as the 1750s, but now horse and cattle raising increased rapidly in importance to provide commodities for a market upon which many Indians had become dependent to supply their basic material needs. Whereas freely available deer had once benefited all, only the individual owners profited from domestic livestock (White 1983:110). The relative egalitarianism of the eighteenth century disappeared as members of certain households and lineages accumulated considerable personal wealth (Woodburn 1982). Typically, a large proportion of these newly affluent individuals were of mixed descent (*métis*), the offspring of European traders who had lived among the Cherokees and Creeks. The decline in egalitarianism coincides with a marked increase in the incidence of padlocks and locking trunks on trade lists and in archaeological contexts dating to that period, reflecting a heightened concern for maintaining personal property (Waselkov 1992).

Cherokee and Creek towns had long been internally dispersed, with household gardens interspersed between widely scattered house clusters (e.g., Bartram 1791:350, 1853:55; Doolittle 1992:397; Schroedl 1986:273–88; cf. Emerson 1992) (figure 9.5). Ongoing excavations of the Creek town site of Fusihatchee, which was occupied from about 1640 to 1814, clearly indicate this settlement pattern (Waselkov et al. 1990:6–22, 39–76). Households were always widely separated within the town, whether one considers seventeenth-century household clusters (each consisting of a rectangular "summer house" and an octagonal

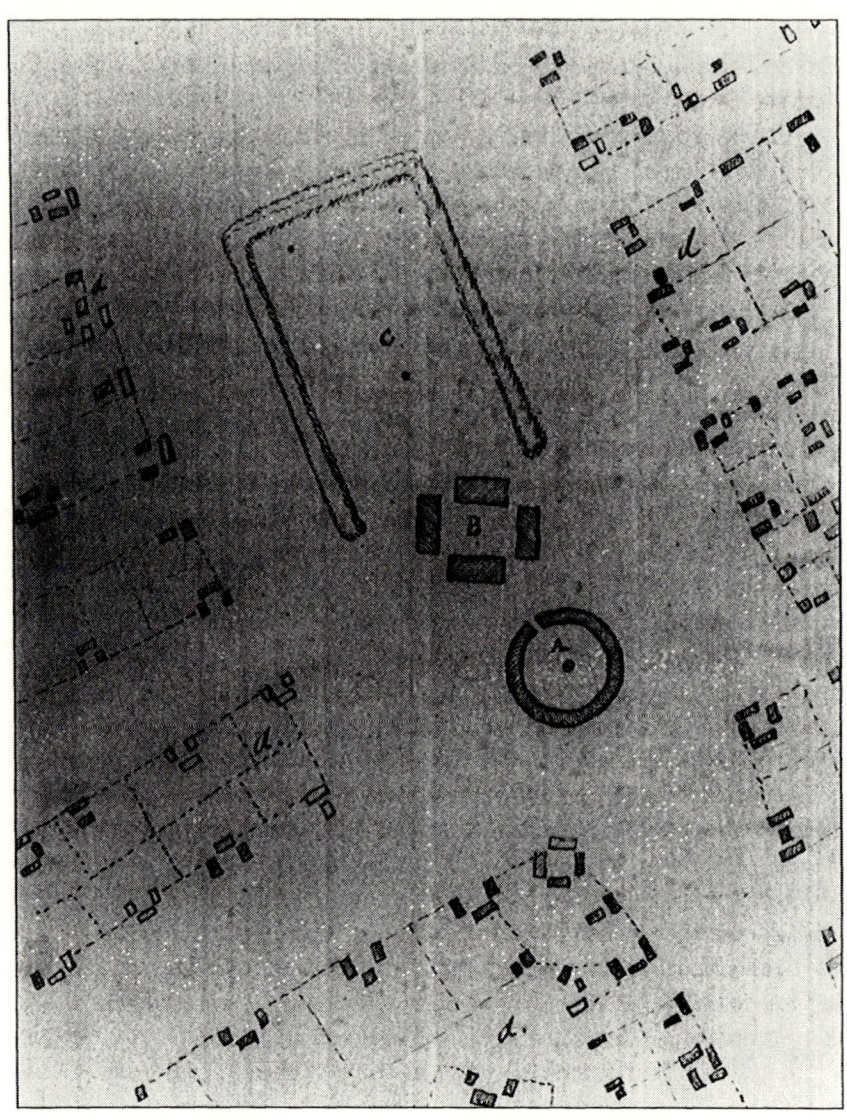

Figure 9.5. Edwin H. Davis's copy of William Bartram's drawing, "A Plan of the Muscogulge or Upper Creek Town," 1789, which shows clusters of Creek domestic buildings and associated garden enclosures arranged around (A) the winter council house, (B) summer square ground, and (C) chunky yard (Smithsonian Institution, National Anthropological Archives, Manuscript #173,683).

"winter house") or eighteenth-century household clusters (made up of multiple summer houses). As dwellings were abandoned and rebuilt in new locations over time, abandoned winter house sites seem to have been reserved for gardens. They would have served admirably, since these semisubterranean "pit houses" were filled with midden upon abandonment, creating nutrient-rich locales suitable for garden plots. Some compelling evidence that gardens were maintained on the locations of abandoned winter houses is found in the settlement data. Whereas the sites of protohistoric summer houses were frequently reused for eighteenth-century summer houses, none of the protohistoric winter house sites was reoccupied by later domestic dwellings. However, the pattern of in-field gardens and out-field bottomland cultivation soon changed as the eighteenth century came to an end.

With the sudden increase in livestock raising (particularly of cattle) beginning in the 1790s, villagers rapidly adopted an isolated farmstead settlement pattern, leaving most former villages occupied only by public structures. A study of land surveys and property evaluations done in the Cherokee area of Georgia in the 1830s demonstrates that, by then, these numerous isolated farmsteads were predominantly located on upland terraces and hillsides, away from the river floodplains that had been favored field locations only a few years earlier (Pillsbury 1983; Wilms 1974). Individual families could find grazing range in the woods around their farmsteads, where they continued to maintain gardens, but they had abandoned the large communal lineage fields that had been the mainstay of Cherokee agriculture.

The floodplains continued to be farmed, however. Evidently the wealthiest and most influential families gained control of the bottomland fields as less powerful families sought private grazing lands in the uplands. Plows and black slaves, which formerly had been rarely seen among the southeastern Indians, soared in numbers, as *métis* families acquired large plantations (generally greater than 50 acres in extent) but lost access to the cooperative labor once supplied by the matrilineages (Pillsbury 1983:62, 67) (figure 9.6).[2]

A nearly identical transition took place among the Creeks in Alabama and Georgia and among the Seminoles of northern Florida (Dickinson and Wayne 1985; Waselkov 1981). Benjamin Hawkins's descriptions of Upper Creek field locations in the Tallapoosa River Valley from the late 1790s indicate that fields occupied the bottomlands, mostly flood-prone silts and silt loam soils (figure 9.7). At that time the

Figure 9.6. This painting by an unknown artist, which is tentatively entitled *Benjamin Hawkins and the Creek Indians* (c. 1800–1810), allegorically depicts the presumed benefits of plow agriculture bestowed on the southeastern Indians by white America (Greenville County Museum of Art, Greenville, South Carolina; Gift of The Museum Association, Inc., with funds donated by Ernst and Young; Fluor Daniel; Mr. and Mrs. Alester G. Furman III; Mr. and Mrs. Dexter Hagy; Thomas P. Hartness; Mr. and Mrs. E. Erwin Maddrey II; Mary M. Pearce; Mr. and Mrs. John D. Pellett, Jr.; Mr. W. Thomas Smith; Mr. and Mrs. Edward H. Stall; Eleanor and Irv Welling; Museum Antiques Shows 1989, 1990, 1991; Collectors' Group 1990, 1991).

Creeks had "lately begun to settle out," according to Hawkins, "for the advantage of wood and raising stock" (Grant 1980:20). By the time Creek lands were surveyed in 1832, preparatory to the federal effort at Indian removal from the Southeast, the bulk of the population was widely dispersed and many council houses stood alone, with not even a remnant of the aggregated households that had once characterized Creek *talwas* (Usner 1985:305) (figures 9.8 through 9.11).

Undoubtedly, the ever-expanding searches for firewood to which Hawkins alluded must have occurred around long-established villages, and other centrifugal forces accelerated this process of settlement dispersal among the Creeks and Cherokees. However, forces such as these

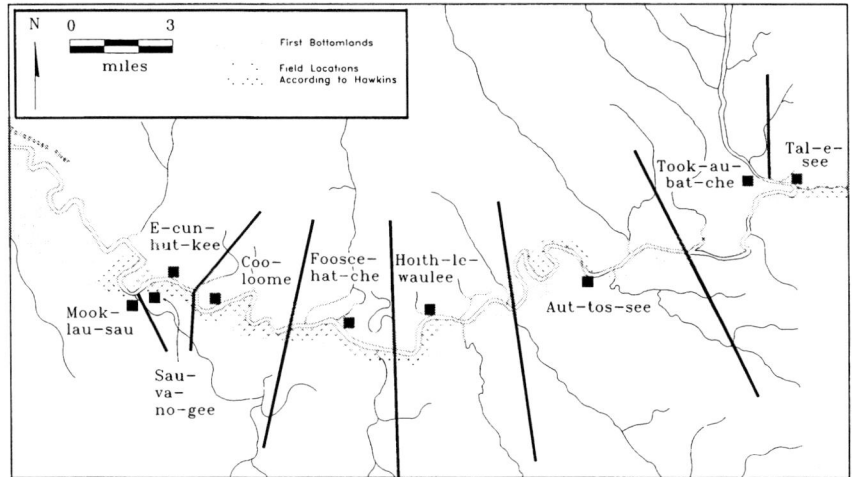

Figure 9.7. Creek Indian agricultural fields in the lower Tallapoosa River Valley in 1798–1799, based on the journal of Benjamin Hawkins, correlated with the locations of silt and silt loam soils (Waselkov 1981:fig. 2).

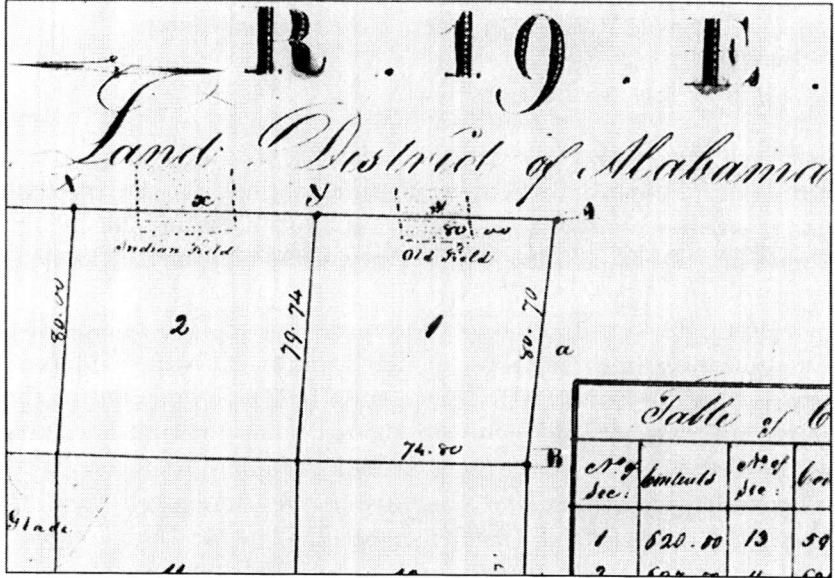

Figure 9.8. Land survey plat showing an "Old Field" and an "Indian Field" (U.S. Land Office, Cahaba Land District, Plat Book 2, 1821–1827, p. 120, T16N R19E, Sections 1 and 2; in the Alabama Department of Archives and History, Montgomery).

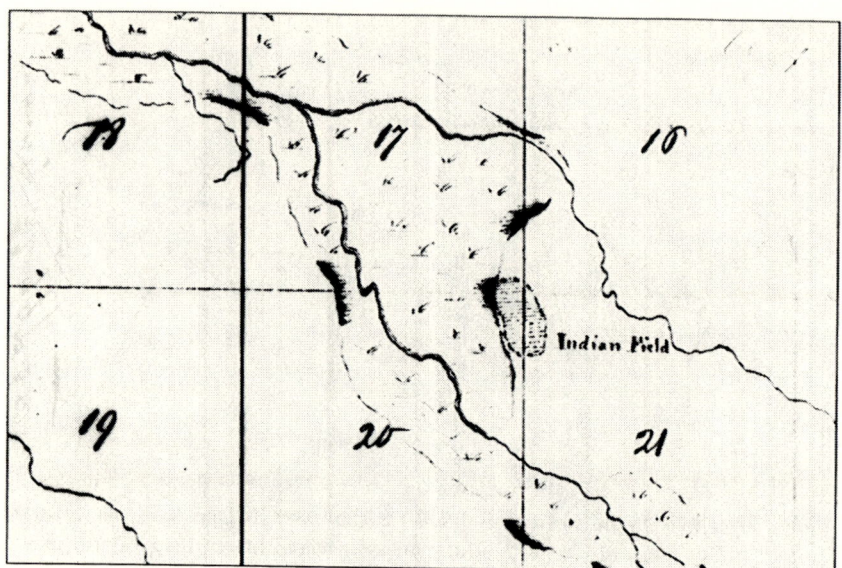

Figure 9.9. Land survey plat showing an "Indian Field" (U.S. Land Office, Cahaba Land District, Plat Book 2, 1821–1827, p. 133, T14N R20E, Sections 16, 17, 20, and 21; in the Alabama Department of Archives and History, Montgomery).

had been effectively dealt with for centuries without leading to the dissolution of traditional settlement patterns. Only when their economic stability was threatened by the loss of the deerskin trade did the Creeks and Cherokees adopt radically new settlement and field location strategies.

One highly significant repercussion was the rapid loss of authority and control over agricultural production by the matrilineages. European traders who had married into southeastern Indian societies during the eighteenth century had long fought the traditional matrilineal control of property and inheritance. Now, as individual families moved apart to provide grazing lands for their livestock, matrilineages were no longer able to muster the labor required to cultivate large common fields. Individual family gardens increased somewhat in size to accommodate each family's needs.

In other words, individual autonomy increased markedly during the early nineteenth century among Creeks and Cherokees. Sharing of labor and resources within lineages declined sharply as the nuclear family

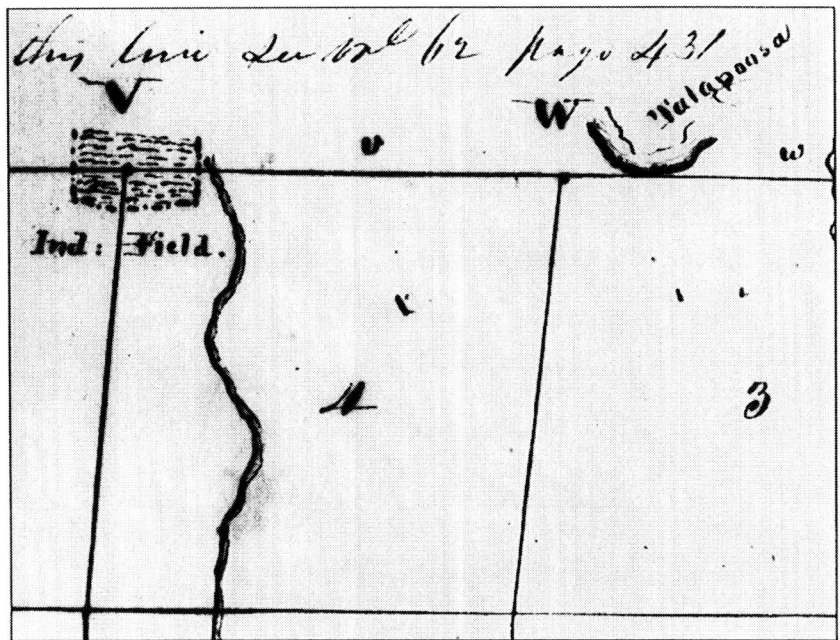

Figure 9.10. Land survey plat showing an "Indian Field" near the Tallapoosa River (U.S. Land Office, Cahaba Land District, Plat Book 2, 1821–1827, p. 136, T16N R20E, Sections 4 and 5; in the Alabama Department of Archives and History, Montgomery).

assumed the role of principal economic unit. This trend corresponds to a more general pattern found among many societies with economies based largely on reciprocity that become increasingly dependent on an external market economy. Anthropologists have adequately documented this phenomenon among foraging bands of fur trappers in the Subarctic (Leacock 1954; Steward 1936), where a number of material and social factors accompanying the arrival of European fur traders in the region led to the appearance of economically autonomous families with exclusive rights to individual tracts of land. This process has since been documented among many other foraging societies (Gardner 1991:544–45, 566), but it is possible to characterize it in even more general terms. Whether the participants were band-level foragers or were segmented or ranked or stratified agriculturists, increased individual autonomy was a frequent adaptive response to the introduction and acceptance of a market economy in a frontier context. The

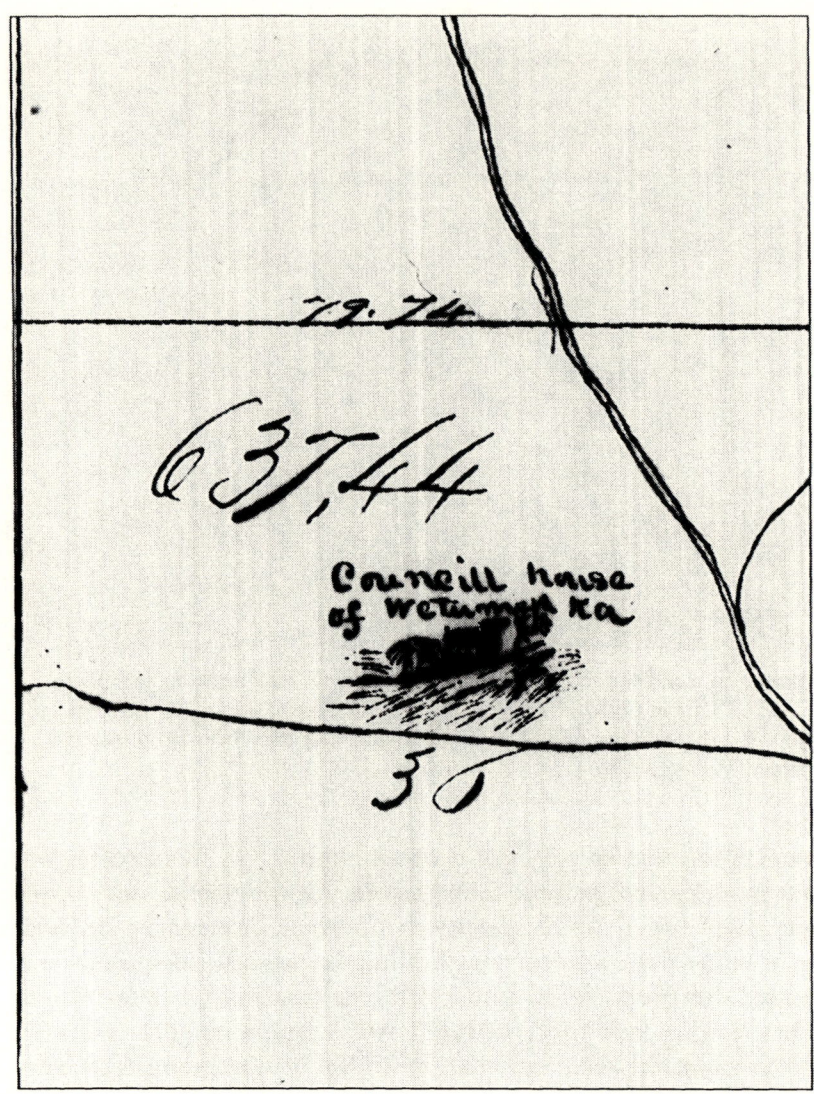

Figure 9.11. Land survey plat showing the "Councill house of Wetumpka" standing isolated from its dispersed population (U.S. Land Office, Tallapoosa Land District, Plat Book 1, 1834, T18N R28E, Section 36; Alabama Department of Archives and History, Montgomery).

Cherokees and Creeks differed from many smaller eastern tribes because they managed to delay for a century this change in social relations, as exemplified by the transition in field location strategies.[3]

Acknowledgments

My thoughts on this topic have benefited from conversations with Tom Hatley, John T. Juricek, Craig T. Sheldon, Jr., Peter H. Wood, and Richard A. Yarnell. Archaeological data on Upper Creek settlement patterns were obtained through grants from the National Park Service (administered by the Alabama Historical Commission) and the National Science Foundation (grants BNS-8718934 and BNS-8907700).

Notes

1. A map of Fort Tombecbe, on the Tombigbee River, dated 1765, shows extensive Choctaw Indian corn fields in the bottomlands across the river from the fort ("Plan of Fort Tombeckbe, 1765," Crown Collection, Series 3, No. 89, William L. Clements Library, University of Michigan, Ann Arbor). This probing exception suggests that the Choctaws favored field locations in bottomlands when available; apart from the Tombigbee, however, most of the river valleys in their eighteenth-century homeland do not have extensive, fertile bottomlands.

2. The painting in figure 9.6 has been the subject of considerable speculation concerning its origin and content. It has been published several times under the title "William Augustus Bowles at His Trading Post on the Chattahootchee River in Florida, with Members of the Creek Nation" (Poesch 1988:175; Simmons 1983:42), although just when and how it acquired this title and by whom it was painted are unknown. The white man in the painting is definitely not the adventurer and would-be filibuster William Augustus Bowles, whose appearance is known from a well-documented portrait (Henri 1986:155; Wright 1967:frontispiece). He may, however, be Benjamin Hawkins, who was federal agent to the Creeks from 1796 to 1816; he at least resembles the painted and engraved likenesses of Hawkins (Henri 1986:153; Pound 1951:frontispiece). Some have suggested that the setting depicts Hawkins's Creek agency on the Flint River (John Juricek, personal communication 1994), although a plan of the agency drawn in 1810 does not match the painting in many details (Mauelshagen and Davis 1969:38–39; Wright 1986:144). In fact, the church steeple and ships under sail in the far left background suggest a coastal location rather than the Creek country. One further confusing element of the painting is the Indian man centrally portrayed, very authentically, in beaded moccasins and leggings, white cloth shirt, and red cloak, with silver arm and wrist bands and a plumed turban. William Augustus Bowles, self-styled "Di-

rector General of the Creek Nation," was painted in 1791 wearing a nearly identical outfit (Wright 1967:51).

Whoever the cast of characters turns out to be, the painting's principal interest derives primarily from the complexity of its symbolic imagery. As an allegorical representation of the plan of the early federal period government to introduce the Indians to "civilization," this work of art is unequaled. White America offers the plow to humbly grateful southeastern Indian men, enabling them to reap abundant harvests and gain other benefits of European culture, as represented by the prosperously pastoral background scene. Of course, Benjamin Hawkins was the foremost promoter of the civilizing policy in the Creek country during this period.

3. That there were alternating cycles of emphasis on upland and bottomland plant exploitation extending deep into the prehistoric past of eastern North America has been suggested by Richard A. Yarnell (personal communication 1992): "*Significant* crop plant remains come from sites in various locations during the latter part of the Late Archaic, mostly from upland sites during the Early Woodland, mostly from lowland sites during the Middle Woodland, largely from upland sites during the Late Woodland (A.D. 300–800), and largely from lowland sites during the Mississippian period (A.D. 1000–1400). This pattern is especially marked in the American Bottom region [of the central Mississippi River Valley], where most of the Late Woodland sites are in the uplands."

CHAPTER 10

Interregional Patterns of Land Use and Plant Management in Native North America

JULIA E. HAMMETT

The landscape of North American archaeology has been dominated by regional perspectives, chronologies, and cultural and environmental reconstructions. Lack of interregional comparisons has hampered our understanding of pan–North American developments and events. By comparing land-use strategies across regional boundaries, several broad patterns emerge. They encompass basic structural dynamics of the landscape, important economic plant families, and key interregional events marked temporally by the expansion of two exotic cultigens, corn (*Zea mays* ssp. *mays*) and beans (*Phaseolus vulgaris*).

Let me begin by admitting that these observations were originally inspired by a variant of Dr. Yarnell's obligatory doctoral written exam question. As his only student of western North America, my special task was to compare the current eastern paleoethnobotanical data base with other, less well understood, western North American data. In the process of carrying out this task, I discovered several broad-based trends. These observed trends are tentative because of scant coverage and the limited availability of regional overviews. Of course there are various regionally and locally important plant resources not included in this survey, but my purpose is to focus attention upon patterns of plant use common to eastern and western North America.

Landscape

Our best and earliest accounts of the native North American landscape are derived from the eastern seaboard. The map illustrated in fig-

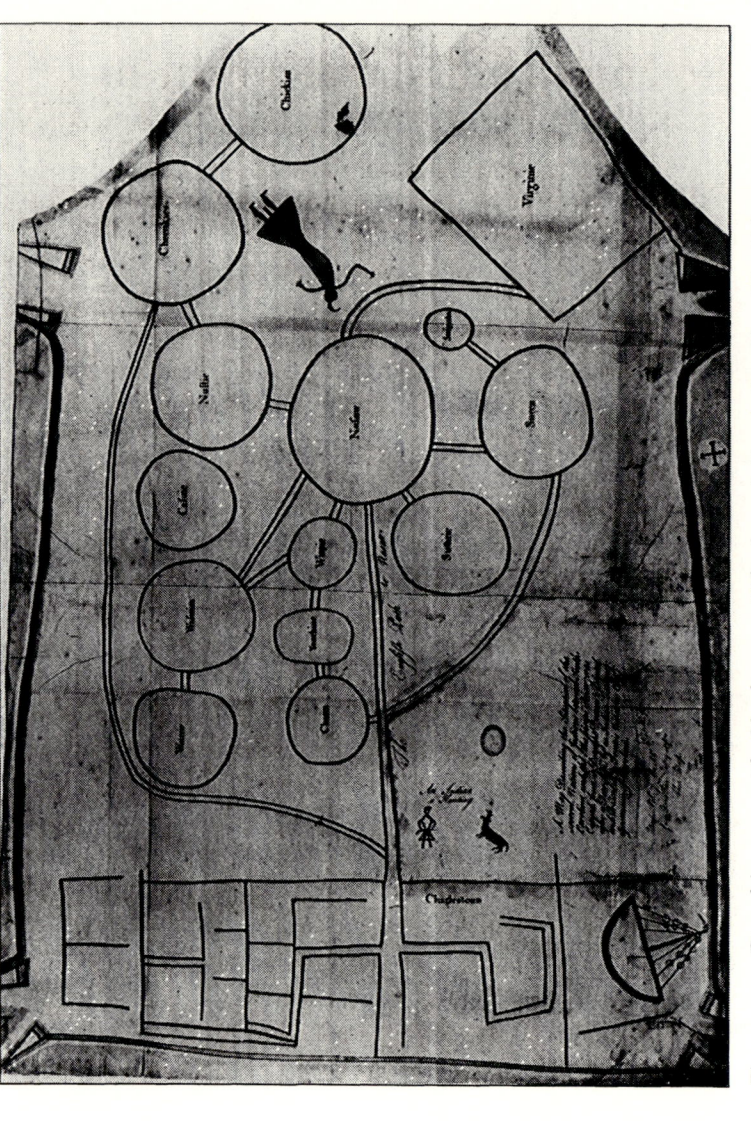

Figure 10.1. Deerskin map with Charlestown (Charleston), South Carolina (English copy, c. 1721. Colonial Office Library 700, North American Colonies, General No. 6[1], Public Records Office, Kew, London. Photographic duplicate provided by the North Carolina Collection, University of North Carolina Library at Chapel Hill).

ure 10.1 provides a rare regional perspective from the viewpoint of a Native American leader of the eighteenth century. The inscription on this map specifies it was "Copyed from a Draught Drawn & Painted upon a Deer Skin by an Indian Cacique [chief]." Waselkov (1988) has deduced that this map was made by a *cacique* of the Catawba Confederation and the map in figure 10.2 by a leader of the Chickasaw Nation.

Landscape ecologists (Forman and Godron 1986) have noted the importance of studying landscape structure in terms of three basic features: patches, corridors, and surrounding matrix. On this map cognized *patches* represent nations, towns, and settlements. Trails, paths, rivers, and other transportation arteries are here designated as *corridors*. Along travel and trade corridors, there are some types of protection, although there are some concomitant vulnerabilities. The large amounts of unbounded space on these maps are apparently lands adjacent to and between territories controlled by specific Indian nations. In the southern Great Lakes region, Hickerson (1965) has documented politically acknowledged buffer areas that had the direct effect of insulating various patch types, including settlements, fishing and hunting grounds, and gardens, and several indirect effects related to resource management, conservation, and military defense. Together with patches and corridors, these buffer areas provided the basic structural components of anthropogenic landscapes in the Eastern Woodlands of North America.

The centers of habitation sites in the Eastern Woodlands were typically composed of domestic structures and small garden plots similar to the "dooryard" gardens found throughout much of Latin America today (Chavero and Averez-Buylla Roces 1988; Kimber 1973). Adjacent to the core of the community were fields of crops, interlaced with old fields lying fallow. In these open old-field areas, perennial crops such as berries, fruit and nut trees, and several economically valuable "weeds" typically thrived (Bartram 1973). A wooded zone beyond the fields served as an extremely valuable combination of orchard, hunting park, and wood lot, supplying the local inhabitants with many of their basic needs and a gradual territorial margin. This landscape was maintained through a series of management procedures.

Less well known are indigenous western landscapes, with the exception of Pueblo cultures of the semiarid Southwest. Pueblo landscapes have changed remarkably little since first contacts with Spaniards in the late 1500s. The older parts of Pueblo towns are still characterized by rectangular stone houses situated around courtyards or plazas. Dooryard

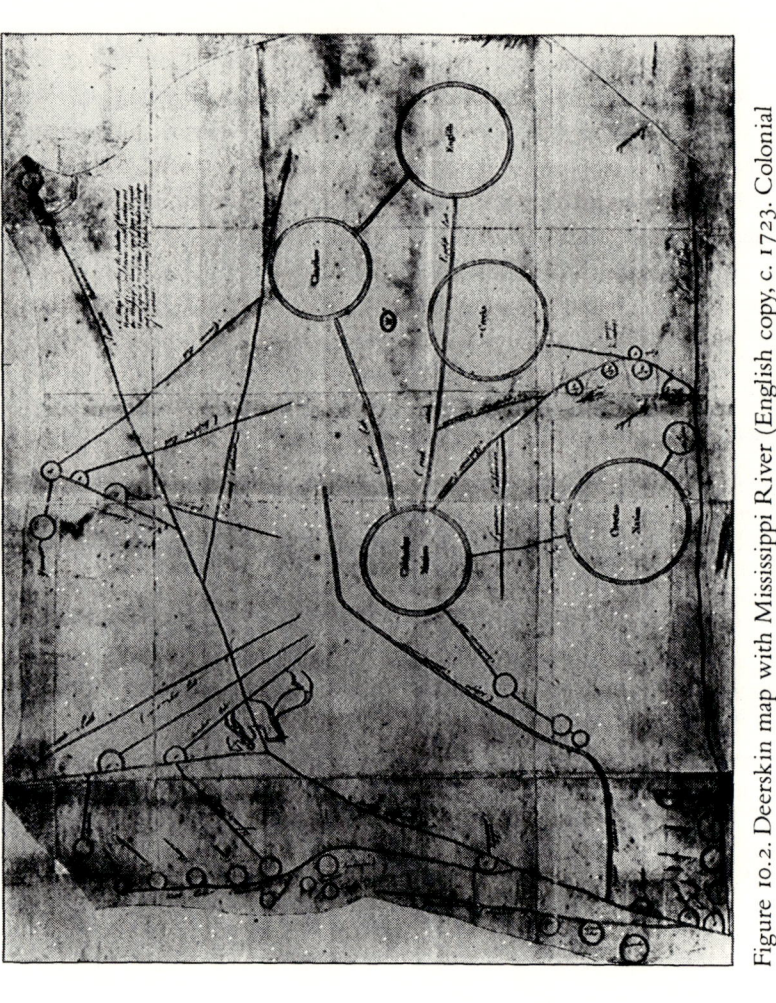

Figure 10.2. Deerskin map with Mississippi River (English copy, c. 1723. Colonial Office Library 700, North American Colonies, General No. 6[2], Public Records Office, Kew, London. Photographic duplicate provided by the North Carolina Collection, University of North Carolina Library at Chapel Hill).

gardens are best known from the Zuni pueblo. Located just beyond the walls of the town and designed for hand irrigation, these gardens are formed by grids of low ridges, prompting the vernacular name "waffle gardens" (Ladd 1979). Farther from towns, along alluvial flats adjacent to rivers and in the deep sands of stabilized dunes, are larger fields of corn, beans, and squashes (*Cucurbita* spp.). Today at their edges are small groves of historically introduced peach (*Prunus persica*) and other fruit trees. Most farming towns of the Southwest are located along major tributaries.

Throughout the nonfarming regions of the West, prehistoric and historic settlements were in close proximity to important resource patches such as swamps, grass fields, and woodlands. Like farming communities, these nonfarming settlements were often located along transportation/exchange corridors such as waterways and land routes or at route junctions such as the foot of mountain passes. Frequently settlements were established where a variety of resource zones coalesced and were situated near water (along drainages and the shores of oceans, lakes, marshes, and springs). Even where farming was not practiced, water was the single most important determinant of settlement pattern. Sites 5,000 years old and older tend to be located at slightly higher elevations than later sites within local catchment/watershed areas. Earlier sites are often on upper terraces, shorelines, small hills, or ridges slightly farther above and back of later sites.

For decades archaeologists have speculated regarding settlement patterns and changing spatial relationships between human beings and their quarry. These discussions are ongoing in every region in North America. Researchers must take a broader landscape perspective and investigate why some of these fundamental geographic relationships are continentwide.

Management

The historical use of fire for wildlife harvesting by Native Americans has been documented in the Southeast (Arber 1910; Bland 1651; Hammett 1992; Lefler 1967), the Northeast (Morton 1637; Russell 1983; van der Donck 1846), California (Cornett 1987; Lewis 1973; Shipek 1989; Timbrook et al. 1982), the Great Basin (Fowler 1986, 1994; Winter and Hogan 1986), and the Subarctic (Lewis 1977; Lewis and Ferguson 1988). In the American Subarctic, Lewis and Ferguson (1988:61) have suggested that the edge areas of fire-managed corridors and patches would have provided native resource managers with a "greater abundance of plant-animal resources and a higher measure of hunting predictability."

We can safely assume Native Americans were aware that the indirect effect of their burning was to increase productivity of the land in terms of food crops, medicinal herbs, weaving materials, and forage for game. This assumption is based not only on observations made by plant and animal ecologists regarding numerous short-term responses of these organisms to fire (Hammett 1991, 1992; Mellars 1976), but on interviews with native experts who practice or remember practicing similar techniques in the Subarctic (Lewis 1977; Lewis and Ferguson 1988), the Great Basin (Fowler 1986, 1993), and southern California (Shipek 1989).

Farming was not practiced in the far western part of the continent. However, even among nonfarming groups in the Great Basin, Northwest Coast, and California, small garden plots were sometimes tended by specialists for herbs, medicines, and basketry materials. Rather than being dooryard gardens, however, these special plots were more often found in discrete locations known only to their caretakers.

Some plants, such as *Rhus trilobata* (lemonade berry) and *Salix* sp. (willow), were manipulated by either pruning or burning in the western Great Basin and southern California. In the case of *Rhus trilobata,* burning has been demonstrated to lengthen and straighten the shoots (Bohrer 1983; Fowler 1986). Coppicing by thorough pruning of both *Rhus trilobata* and *Salix* sp. generates choice shoots for basketry and other weaving industries (Anderson 1993; Fowler 1994). In the case of *Salix* sp., this practice had to be repeated at least annually in order to maintain the prized resource (Fowler 1994). The shoots that resulted from this anthropogenic process have been found in prehistoric split-twig figurines in shelters in the Great Basin and the Southwest (Bohrer 1983). Although data are lacking for similar artifacts elsewhere in North America, we can attribute this to better preservation in drier environments.

Throughout most of temperate North America, important medicinal and spiritual plants, such as tobacco (*Nicotiana* sp.), were managed in small plots by a combination strategy of seed planting, burning, and clearing (Driver 1969; Winter 1991). One medicinal plant, yaupon (*Ilex vomitoria*), a holly shrub, was transplanted in groves and hedges

Figure 10.3. Engraving of John White's drawing of *The Manner of Making Their Boates*. In this scene fire is being used for various stages of boat manufacture. (From *America 1585: The Complete Drawings of John White*, edited by Paul Hulton, © 1984 by The University of North Carolina Press. Used by permission. Photographic duplicate provided by the North Carolina Collection, University of North Carolina Library at Chapel Hill.)

near communities. Remnants of this aboriginal practice are scattered throughout the maritime and coastal plain of the Southeast (Hammett 1992; Merrill 1979).

With few exceptions, management of tree crops was not documented in early historic accounts, except in discussions of general clearing and burning or the use of wood (figure 10.3) and sap products (Munson 1988). Even these accounts are rare. In 1773, William Bartram traveling in Georgia found persimmons, honey locust chickasaw plum, beauty berry, red mulberry, hickory, and black walnut growing in fallow fields adjacent to old Indian settlements. He noted that some "old peach and plumb orchards" appeared "yet thriving and fruitful" (Bartram 1973:343).

Specifically, reports of nut tree management are noticeably lacking in early accounts. This is unfortunate, given the prehistoric importance of nut crops in both eastern and far western North American woodlands (see Gardner, this volume). Bartram noted that (with the exception of peach) these trees "are natives of the forest, yet they thrive better and are more fruitful in cultivated plantations, and the fruit is in great estimation with the present generation of Indians, particularly juglans exalta, commonly called shell-barked hiccory" (Bartram 1973:38). Prescribed burning would have been an unlikely tool for management of hickory trees, as most species are highly susceptible to fire (Deam 1921). However, an opening of the forest canopy by thinning and weeding non-fruit-bearing trees would have resulted in larger, higher-yielding trees (Munson 1986; Yarnell 1964).

In the far West, stands of pinyons (*Pinus* sp.) in the foothills of southern California (Anderson 1993) and the southern Sierra Nevada Mountains were managed (Fowler 1986, 1994). Groves of honey mesquite (*Prosopis glandulosa*) in the Mojave Desert were pruned and cleared (Fowler 1994). Desert fan palms (*Washingtonia filifera*) of oases in southern California were maintained with burning and clearing (Cornett 1987; Shipek 1989). Cahuilla oral history indicates the range of these palms was expanded by human dispersal of their seed (Cornett 1987:17).

No doubt the lack of more substantial management accounts is partly because of the life span of trees, which is often considerably longer than that of human beings. The longevity of trees makes the tracing of mutualistic relationships problematic. Nevertheless, seed planting, transplanting, clearing, and possibly burning would have affected tree populations over generations.

Together these data depict a series of management and harvest strategies utilizing a wide range of plant life. Prescribed burning was probably the most widely used resource management technique. Farming was limited to bottomlands along major rivers in temperate regions, but gardening or the tending of various plots by specialists was widespread.

Upon close scrutiny, we can identify a finite set of plants that were common members of anthropogenic landscapes throughout much of North America. Having illustrated the general structure and dynamics of Native American landscapes, I turn attention to some specific plant families that were economically important throughout the continent and highlight some important mutualistic relationships.

Table 10.1: Important Pan–North American Plant Families[1]

Nut/Fruit Trees and Shrubs	Roots/Tubers/Corms
Ericaceae (Heather)	**Apiaceae** (Parsley)
Fabaceae (Legume)	**Liliaceae** (Lily)
Fagaceae (Oak)	**Typhaceae** (Cattail)
Juglandaceae (Walnut)	*Food/Containers*
Rosaceae (Rose)	**Cucurbitaceae** (Gourd)
Small-Seeded Herbs, Forbs, and Subshrubs	Exotic Foods
Amaranthaceae (Pigweed)	**Fabaceae** (Legume)
Asteraceae (Composite)	**Poaceae** (Grass)
Chenopodiaceae (Goosefoot)	*Medicines*
Portulacaceae (Purslane)	**Solanaceae** (Nightshade)
Poaceae (Grass)	

[1]Common names of plant families are given in parentheses.

All in the Family

Archaeobotanical evidence indicates that about fifteen plant families contain taxa that were economically important to Native Americans throughout the temperate regions of North America (table 10.1). Economic plants in these families tend to thrive on intermediate levels of disturbance such as burning, clearing, and cultivation. Although specific taxa may vary by region, the importance of these families in pan–North American subsistence is clear.

Four selected pairs of plant taxa within three families (Poaceae, Asteraceae, and Juglandaceae) have been mapped (figures 10.4 through 10.7) to compare and contrast geographic distributions with archaeobotanical evidence. The mappings are limited to the southern half of North America because it is there that most archaeobotanical work has been conducted. The mapped regions in figure 10.8 are based on regional floras with consideration of archaeologically perceived regions.

Two tree genera from Juglandaceae, the walnut family, are *Juglans* (figure 10.4, left), containing black walnuts and butternuts, and *Carya* (figure 10.4, right), including pecans and hickory nuts. The maps illustrate that the distribution of *Juglans* is more nearly ubiquitous than that

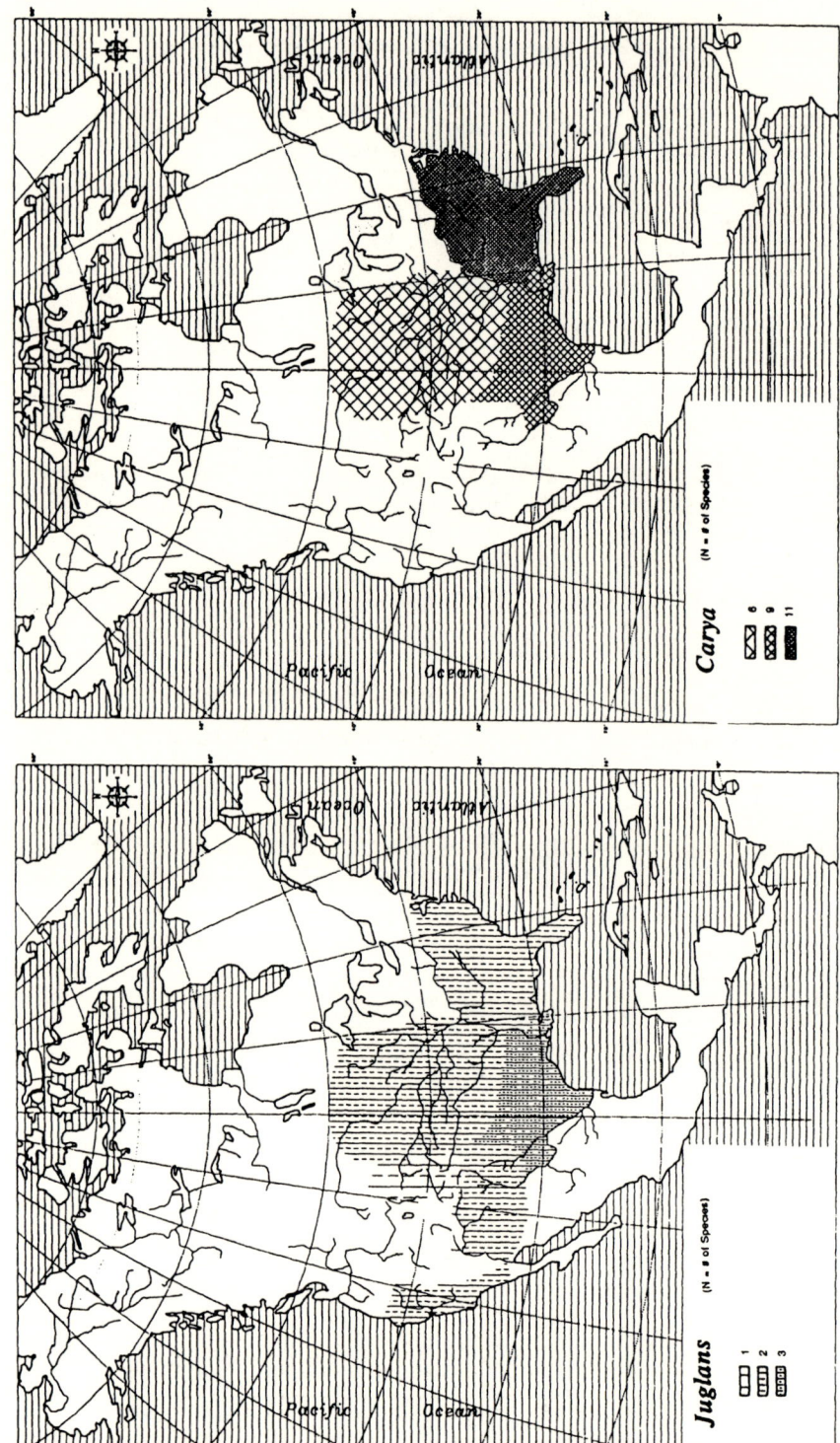

Figure 10.4. Distribution of Juglandaceae family.

of *Carya*. *Carya* is limited to the eastern half of North America, whereas *Juglans* sp. occurs across the temperate regions of the continent.

Two species of this family are native to California, *Juglans californica* and *Juglans hindsii*. The fruits of *J. hindsii* are characteristically 15 mm longer and 10 mm wider than fruits of *J. californica* (figure 10.9). Taxonomically, *J. hindsii* is considered a larger, more treelike variety of *J. californica* (Jepson 1970; Munz and Keck 1973). *Juglans hindsii* is said to be found "about old Indian campsites" in central California (McMinn and Maino 1963:157; Munz and Keck 1973:909). Radiocarbon dating and stratigraphic analyses indicate a strong likelihood that some of the large prehistoric habitation sites in central California were continuously occupied, or at least frequently revisited, for hundreds of years (Moratto 1984). This larger-sized walnut species may have shared a mutualistic relationship with native inhabitants of central California, whereby people selected for larger nuts over a number of generations through seed planting, transplanting, pruning, or merely clearing around larger-fruited plants. However, more direct evidence of this dynamic relationship is lacking, and the association remains for the time circumstantial.

The next two pairs of maps illustrate the geographic distributions of four native North American grain crops, two so-called wild rices (figure 10.5), *Oryzopsis* and *Zizania,* and two other well-known economic grasses, *Hordeum* and *Phalaris* (figure 10.6). As these maps illustrate, the rice grass geographic distributions are rather divergent. Ecologically, *Zizania* occupies wetlands, in particular shallow lakes and slow-moving rivers and streams (Aiken and Darbyshire 1983). *Oryzopsis* inhabits the arid and semiarid terrains characteristic of western North America. Despite separate growing habits and exploitation demands, however, these two genera shared a somewhat similar role in the subsistence bases of Native Americans in different regions. It would be simplistic to make too much of the fact that they bear the same common name of "Indian rice grass." However, this common name invokes an image related to growth pattern: both occur in relatively dense and homogeneous patches within their habitats. Perhaps one also gets an image of a collection strategy with seed beaters and baskets, albeit from a boat for *Zizania.*

The bulk of our archaeobotanical information for these grasses reflects modern geographic distributions with one intriguing possible exception. Two unusual taxa of grass grains were recovered from a 5,000-

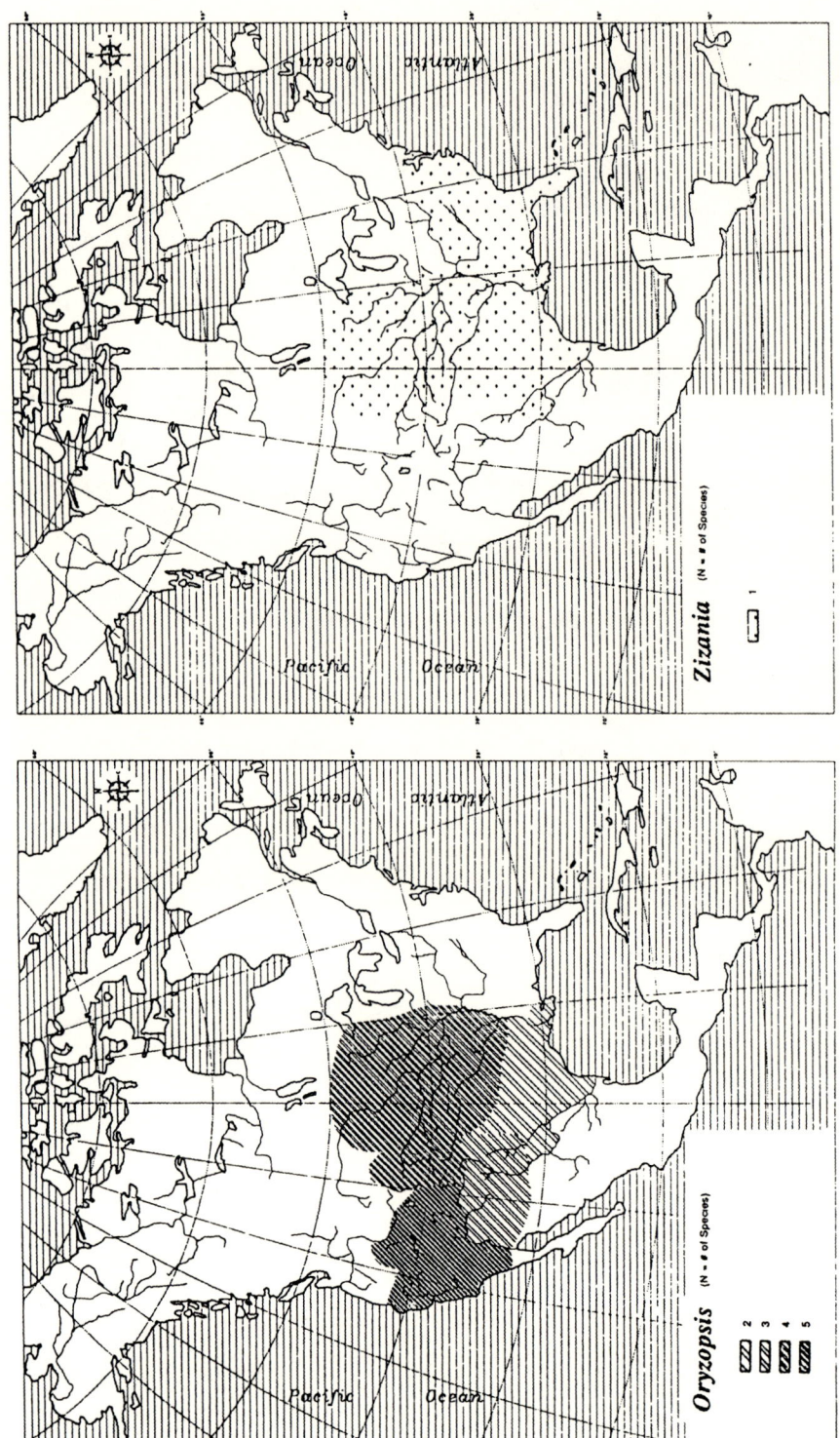

Figure 10.5. Distribution of North American "wild rice" grass genera *Oryzopsis* and *Zizania* (Poaceae).

year-old open shelter site in coastal California (Singer and Atwood 1987). One taxon, with grains averaging more than 6 mm in length, closely resembles *Zizania* (Hammett 1991). The current westernmost range of *Zizania* spp. extends to the south as far as Texas and to the north as far as Manitoba, although it has been reportedly introduced into Saskatchewan (Aiken and Darbyshire 1983). The possibility cannot be precluded that this genus extended into the wetlands of California 5,000 years ago. On the other hand, until more substantial achaeobotanical work has been conducted in California, one must remain skeptical of this single find.

Phalaris and *Hordeum* (figure 10.6) are widespread throughout North America. Yarnell (in press) suggests that Native Americans of the Eastern Woodlands and the Southwest may have both used these cool-season grasses. Archaeobotanically they have been recovered in all the southern regions, but nowhere have researchers found any evidence of morphological change indicating domestication. In California, species of both genera have been identified from coastal sites dating to 1000 B.P. and possibly as early as 5000 B.P. *Phalaris* grains have also been recovered from the Hohokam area in the Southwest in contexts dating 1100–1000 B.P. (Bohrer 1987).

In the eastern United States, *Phalaris caroliniana* (maygrass) first appeared in Tennessee about 3,000 years ago and in Kentucky somewhere between 3,000 and 2,000 years ago. There is some evidence of the cultivation of this species (Cowan 1978a). Overall it appears that *P. caroliniana* was broadly utilized in the Southeast and Midwest and was utilized to a lesser extent in the Southwest. Apparently a greater diversity of *Phalaris* species was harvested in California than elsewhere.

Hordeum pusillum (little barley) has been identified from several prehistoric sites in the Midwest dating to between 2000 and 1000 B.P. (Asch and Asch 1985b; Cowan 1985b; Fritz 1990); from Hohokam sites in the Southwest by 2000 B.P. (Adams 1985; Bohrer 1962, 1970); and from one protohistoric context in southern California (Hammett 1990). Again, the data indicate that a great diversity of *Hordeum* species was utilized in California, rather than the monospecific crop farther to the east.

It is intriguing that the majority of current evidence for the exploitation of *Hordeum pusillum* and *Phalaris caroliniana* exists away from the North American centers of diversity for these two genera. Given

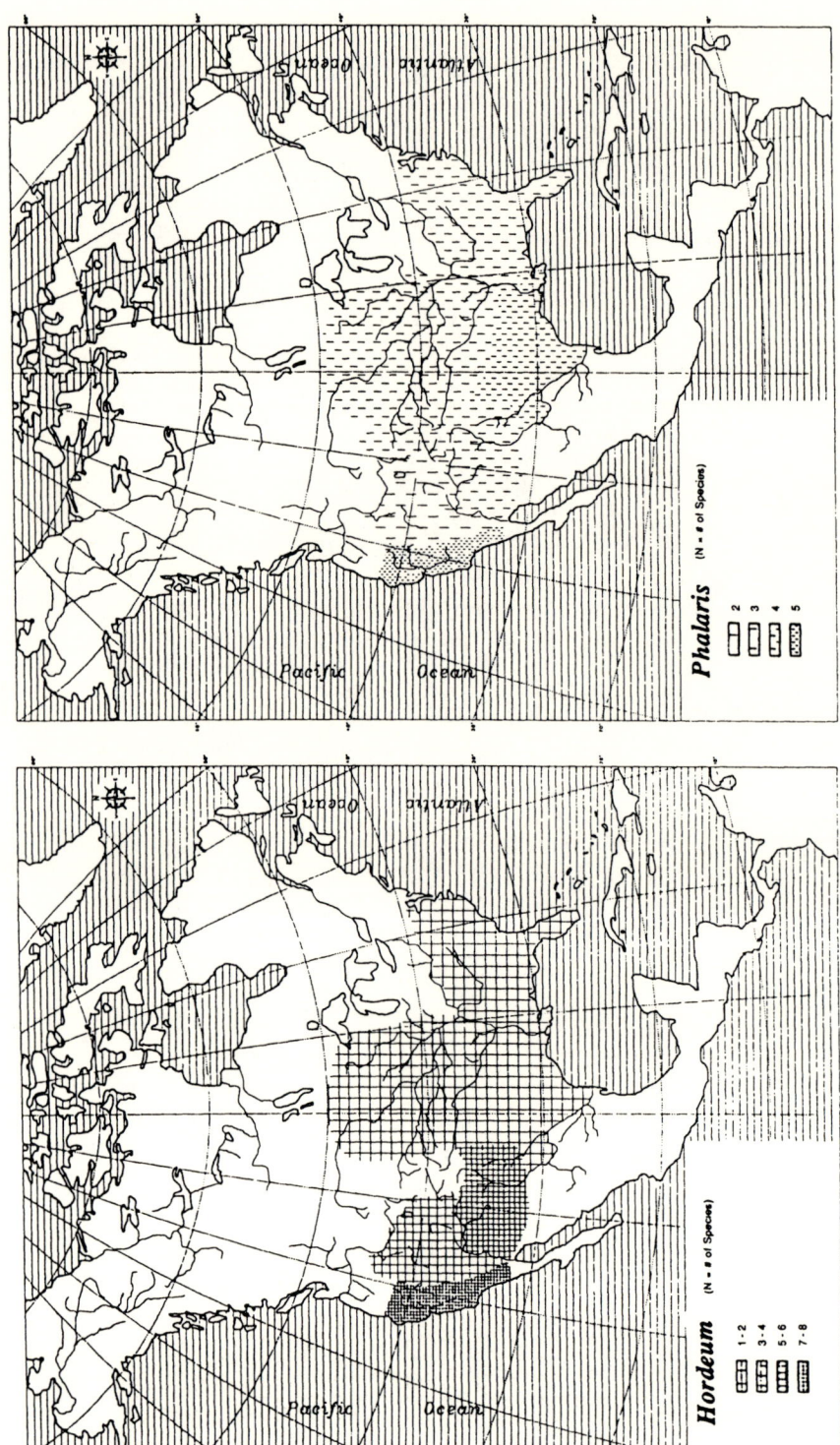

Figure 10.6. Distribution of *Hordeum* and *Phalaris* (Poaceae).

the limited data base, the possibility must be allowed that this is an artifact of differential preservation and differential investigation.

The last pair, *Iva* (sumpweed) and *Helianthus* (sunflower) (figure 10.7), are members of the Asteraceae, or composite family. It is believed that domesticated *Helianthus annuus* is derived from a western stock (Heiser 1954, 1969, 1976), although the current earliest archaeobotanical evidence for a domesticated form is from the Ozarks and Tennessee (Fritz 1990). The earliest evidence for a wild *Helianthus* species in the West is from Cowboy Cave in the Great Basin and was recovered from deposits more than 8,000 years old. This taxon is either *H. anomalus* (Hogan 1980) or *H. petiolaris* (McVickar 1991).

Iva annua (sumpweed), like the *Helianthus* species, occurs throughout southern North America; unlike sunflower it was apparently domesticated in the vicinity of or near where it is native. This is somewhat problematic theoretically because current understanding of the domestication process leads to the belief that isolation of a strain is generally associated with mutualistic relationships that gradually lead to domestication (Rindos 1984). Yet even within its native range, populations of *Iva annua* may have been isolated outside their bottomland microhabitats, for example, by being grown in an upland setting. One intriguing similarity of these two taxa is that sometime after they both appeared in domesticated form, they continued to occur together in sites throughout the Midwest and the Southeast (Fritz 1990). Thus, following Rindos (1984), these two composites apparently shared similar agroecological niches in native North America for more than a thousand years.

Exotic Origins

It would not be appropriate to ignore the battle of indigenous domestication currently under debate among eastern North Americanists, but emphasis on this topic has somewhat sidetracked the hemispheric importance of two exotic crops introduced prehistorically. Corn (*Zea mays*) and beans (*Phaseolus vulgaris*) continue to be important staples to millions of people throughout the world. Sidestepping their origins in South America and Mesoamerica, their paths of utilization in North America proper began in the Southwest.

Corn is arguably the single most important crop of the New World. Perhaps the best-known archaeobotanical "fact" for fifteen years and

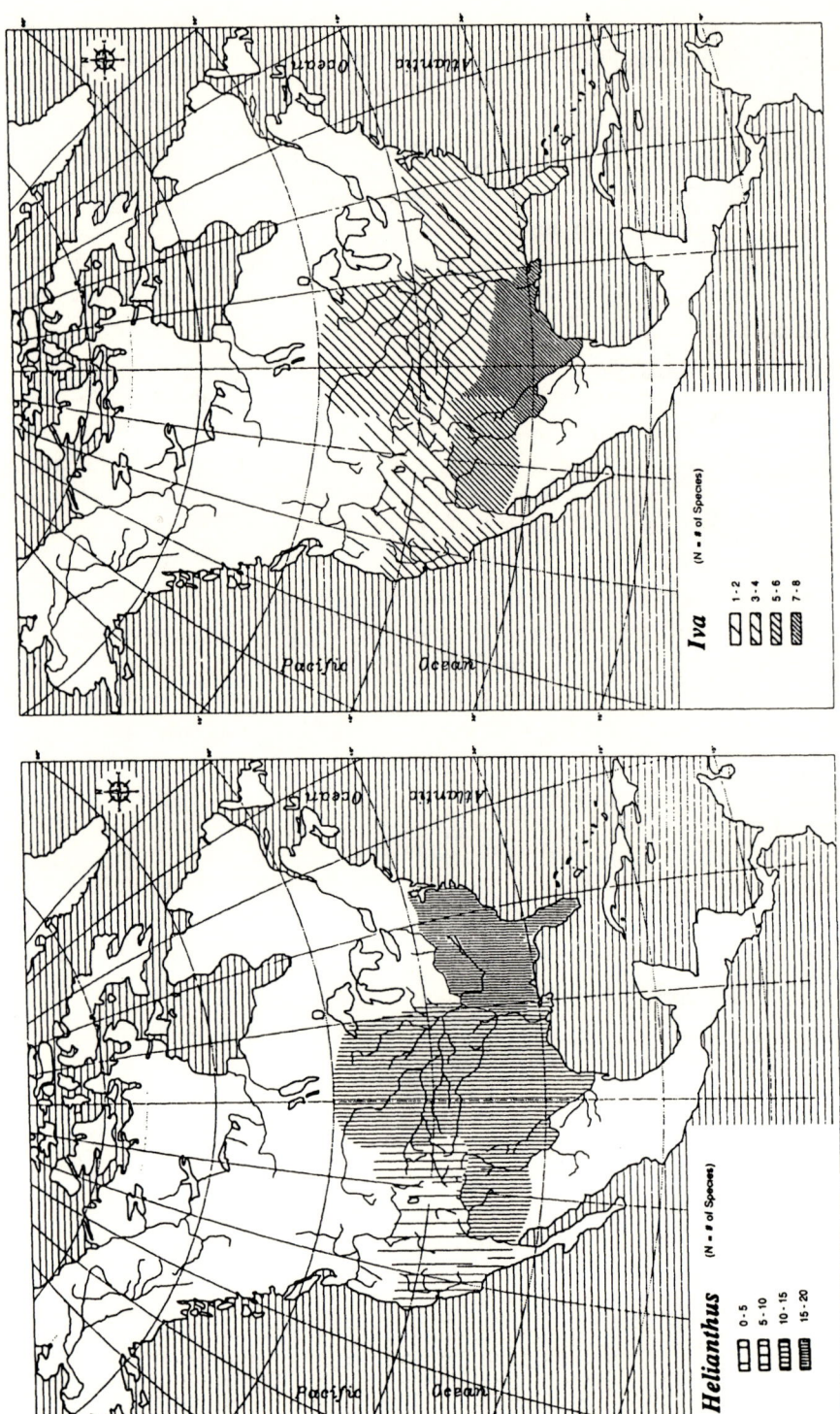

Figure 10.7. Distribution of *Helianthus* and *Iva* (Asteraceae).

still in most anthropological texts is the 2300 B.C. (4250 B.P.) date for corn at Bat Cave in southeastern New Mexico. Recently, more reliable stratigraphic controls have pushed forward Bat Cave's corn evidence to 2500 B.P. (Berry 1985; Wills 1988b). Undaunted researchers currently working in the northern Anasazi frontier along the Colorado and Utah borders are trying to reestablish the earlier date. Conservatively, corn entered North America proper via the Southwest by 3000 to 2000 B.P. (Adams 1994). By the latter date, corn had appeared north of the Southwest.

Between 2100 and 1700 B.P. corn expanded northwestward into the Great Basin (Wilde and Newman 1989) and northeastward into the Great Plains (Adair 1988) and the Eastern Woodlands (Chapman and Crites 1987; Riley et al. 1994). Over the next thousand years corn gradually became well established in the Southwest and eventually became a staple in much of that area, but during most of this time it remained relatively rare throughout much of the Southeast (Fritz 1990), the Great Plains (Adair 1988), and the Great Basin (Winter 1973, 1974, 1976, 1982; Wilde and Newman 1989). Then, about 1,200 years ago, there was a dramatic increase in the presence of corn at the northern and eastern perimeters of the Southwest. Over the next 200 to 300 years, this apparent subsistence shift was repeated in the Great Basin (Adams 1994), the Great Plains (Adair 1988), and the Southeast (Fritz 1990).

At the same time that consumption of corn was intensifying in the regions north of the Southwest, cultivated beans (*Phaseolus vulgaris*) were making their first appearances in these regions (ca. 1200–850 B.P.). Like corn, beans traveled simultaneously into the Great Basin (Berry 1972; Dodd 1982), the Great Plains (Adair 1988), and the Southeast (Yarnell 1986). Again, over the next 200 to 300 years, this crop increased substantially in some locales, although in others beans remained rare or totally absent.

Surely the temporal correlations of the expansion of corn out of the Southwest into such a broad area of North America and then its intensification about a thousand years later with the concomitant arrival of beans into these areas do not mark coincidental events. Rather, we should look to macroscale important social and environmental correlations.

Most regional archaeological chronologies begin with a Paleoindian period ending about 8,000 years ago, followed by a tripartite (early, middle, late) set of prehistoric periods ended by the protohistoric

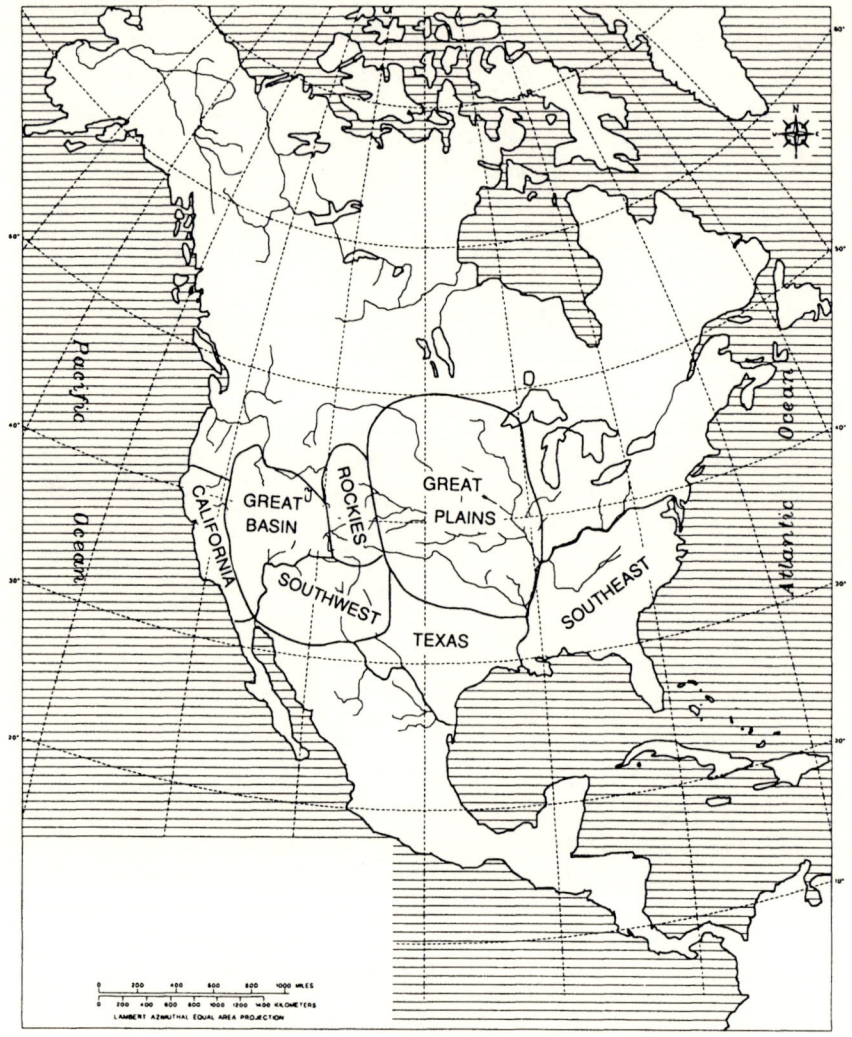

Figure 10.8. Plant geographic regions, based on Britton and Brown 1896;
Correll and Johnston 1970; Hitchcock 1971; Hitchcock and Cronquist 1973;
Jepson 1970; Lakela 1965; Looman and Best 1987; Martin and Hutchins 1981;
Munz and Keck 1973; Porsild and Cody 1979; Radford et al. 1968; Swallen
1940; Weber and Willmann 1992; and Welsh et al. 1987.

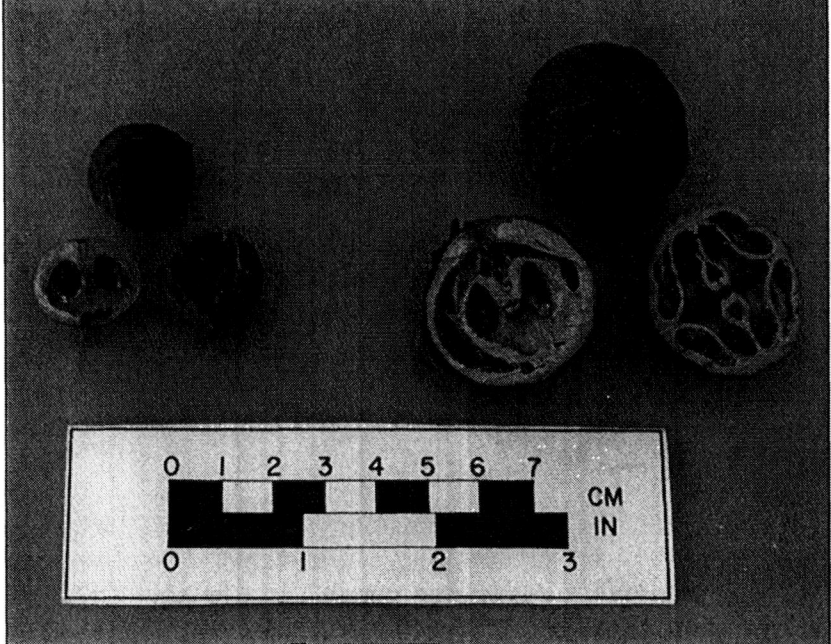

Figure 10.9. Comparison of California-native walnut shells: *(left) Juglans californica; (right) Juglans hindsii.*

period, which marked direct and indirect contacts with European explorers and colonists. The period between 1000 and 800 B.P. has been well established as the transition zone between cultural horizons appearing across the continent. Somewhere between 1000 and 800 B.P. virtually all regional chronologies enter their late prehistoric horizons.

In western North America, the years between 1250 and 650 B.P. were a time of rather severe environmental perturbations. In the northern Southwest, the period between 1250 and 900 B.P. was a relatively prolonged phase of high temporal climatic variability, that is, rapid oscillations in precipitation, probably necessitating the acquisition and storage of surplus food (Dean et al. 1985). From 900 to 650 B.P. oscillations were much lower. From the insider's view this would increase the ability to predict weather patterns; however, when droughts did occur, such as the Great Drought that began in 675 B.P. and lasted twenty-five years (Dean et al. 1985), there would be no relief. In neighboring southern California, Arnold (1992) has suggested that the dramatic increase in political and socioeconomic complexity for the Coastal Chumash that began approximately 800 years ago (King 1990) may have been related to Southern Oscillation phasing within the El Niño/La Niña climatic cycle.

Recent climatological studies have demonstrated that weather patterns in the Southwest, the Southeast, southern California, and much of the Great Basin are all tied to the extreme phases of the Southern Oscillation phenomenon (Gunn 1991; Swetnam and Betancourt 1990). Other regions are less well understood climatologically, but archaeologists across the continent have recognized the occurrence of social developments, particularly around 1000 B.P. Thus it is obvious that key economic changes were tied in part to pan–North American cultural and perhaps environmental mechanisms that crosscut regional boundaries.

Conclusions

Before concluding, I must stress that these macroscale patterns are far from universal. Recently Fritz (1990) demonstrated that multiple evolutionary pathways were pursued in eastern North America. Similarly, western North Americanists appreciate that many groups did not follow socioeconomic paths of development that led to domestication. These communities were probably aware of groups that did, however, and may have experimented, or even deliberately chosen not to indulge, once these paths were well established by their regional neighbors. It is also true that not all plant genera, or even families, occur in all regions. Nevertheless, some similarities in utilized plant taxa and methods are striking among a number of regions.

By the time of contact between European and Native American populations, many native North American landscapes were regularly manipulated with prescribed burning and other disturbance activities. Historic and ethnographic records in the Subarctic and in California depict the opening of grasslands and meadowlands for hunting and the collection of small seeds. Adjacent to these openings were woody shrublands supplying berries and fruits. Beyond these, the local environment allowing, were forests of fuel and building materials and nut trees. This pattern was slightly different in more arid parts of western North America where trees are restricted by water supply; there, trees concentrate along water channels or, for conifers, at higher elevations.

Occupied areas of the Eastern Woodlands resembled open parklands with nut and fruit trees surrounding clearings of old and new fields, with settlements at their cores. Forested buffer areas apparently served as political and economic insulation between recognized groups and leaders. Throughout North America major trade arteries, or corridors, brought the exchange of resources and information, apparently sometimes moving simultaneously in several directions.

Additional archaeological data are necessary to determine the longevity of these landscape and land-use patterns. In general, however, the geographic and archaeobotanical distributions of the selective taxa discussed above indicate broad similarities in subsistence across North America. The exploitation of grains and other small seeds was widespread and significant throughout the region. Mixed subsistence strategies combining nutmeat and small seed harvesting techniques were utilized in the southern woodlands of both the eastern and far western portions of the continent. Archaeobotanical evidence from several regions indicates that grasslands and parklands of nut and/or fruit trees, often skirted with understories of fruit-bearing shrubs, were widespread and readily exploited continentwide over the past several thousand years. Overall, nonfarming landscape patterns exhibit the greatest similarity in the forests of the continental extremes; farming was characteristic of a more restricted region, in the Southwest and eastward. The use of burning and other clearing technologies was common throughout, although specifics varied considerably.

As archaeologists, we understand that our data are biased by our limited identification technology, the processes of differential preservation and taphonomy, and the diversity of continental physiographic zones. For example, little mention has been made of the importance of root and tuber crops, although histories and ethnographies emphasize their importance throughout North and South America. Currently, however, archaeobotanical techniques for obtaining this information are extremely limited.

Unfortunately, our data are further weakened by limited information exchange between scholars in some regions and a paucity of fieldwork in others. These biases, coupled with an absence of interregional perspectives, diminish our potential as North Americanists. Like the other patterns discussed in this chapter, these final observations are probably a result of both environmental and social factors. Existing data indicate that, unlike us, indigenous populations had finer-grained geographic coverage and well-established communication networks; traditionally, information appears to have been readily transferred across the land.

This exploration of North American patterns has identified three directions for further research. A survey of traditional resource-management strategies has revealed a suite of techniques found throughout indigenous North America. Several economically important plant families are ubiquitous in temperate North America, and some of the relationships between human beings and these plant families are shared among regions. Finally, temporal correlations for the introductions and expansions of exotic crop plants transcend regional boundaries, apparently signifying important milestones in pan–North American prehistory. Together these data indicate the importance of information exchange throughout North America and the necessity of linking data between regions in order to comprehend fully the important cultural developments within regions.

These interregional patterns are in much need of refinement. It is hoped that continued research will use, revise, and denounce the observations outlined here. That will be the exciting task of the future, thanks to the solid foundation established over the past thirty years.

Acknowledgments

A number of scholars contributed substantive information and constructive criticism during the drafting of this chapter. They include Karen Adams, James A. Carter, Catherine Fowler, Gayle Fritz, Kristen Gremillion, Joel Gunn, Andrea Hunter, Lisa Klug, Patty Jo Watson, Joseph Winter, Lee Newsom, and one anonymous reviewer. Mapped distributions of plant taxa were computer drafted by Roman Fojud.

REFERENCES CITED

Adair, Mary J.

1988   *Prehistoric Agriculture in the Central Plains.* Publications in Anthropology No. 16. University of Kansas Press, Lawrence, Kansas.

Adams, Catherine F.

1975   *Nutritive Value of American Foods in Common Units.* Agriculture Handbook No. 456. Agricultural Research Service, United States Department of Agriculture, Washington, D.C.

Adams, Karen R.

1985   Little Barley [*Hordeum pusillum* Nutt.] as a Possible New World Domesticate. Appendix 9C in *La Ciudad: Specialized Studies in the Economy, Environment and Culture of La Ciudad,* edited by J. E. Kisselburg, G. E. Rice, and B. L. Shears, pp. 203–37. Anthropological Field Studies No. 20. Arizona State University Office of Cultural Resource Management, Tempe.

1994   A Regional Synthesis of *Zea mays* in the Prehistoric American Southwest. In *Corn and Culture in the Prehistoric New World,* edited by Sissel Johannessen and Christine Hastorf, pp. 273–302. Westview Press, Boulder, Colorado.

Aiken, S. G., and S. J. Darbyshire

1983   *Grass Genera of Western Canadian Cattle Rangelands.* Canada Monograph No. 29. Biosystems Research Institute, Research Branch Agriculture, Ottawa.

Anderson, David G., and Glen T. Hanson

1988   Early Archaic Settlement in the Southeastern United States: A Case Study from the Savannah River Valley. *American Antiquity* 53:262–86.

Anderson, David G., and Joseph Schuldenrein

1983   Early Archaic Settlement on the Southeastern Atlantic Slope: A View from the Rucker's Bottom Site, Elbert County, Georgia. *North American Archaeologist* 4:177–210.

Anderson, E.

1952   *Plants, Man, and Life.* Little, Brown and Co., Boston.

Anderson, Kat
1993   Native Californians as Ancient and Contemporary Culti-
            vators. In *Before the Wilderness: Environmental Management
            by Native Californians,* edited by Thomas C. Blackburn
            and Kat Anderson, pp. 151–74. Ballena Press, Menlo Park,
            California.

Anderson, P. C., editor
1992   *Préhistoire de l'agriculture: nouvelles approaches experimentales
            et ethnographiques.* Monographie du Centre du Recherches
            Archeologique No. 6. Editions CNRS, Paris.

Andres, Thomas C.
1987   *Cucurbita fraterna,* the Closest Wild Relative and Progeni-
            tor of *C. pepo. Cucurbit Genetics Cooperative Report* 10:69–71.

Arber, E.
1910   *Travels and Works of Captain John Smith.* 2 vols. John
            Grant, Edinburgh.

Armelagos, George J.
1990   Health and Disease in Prehistoric Populations in Transi-
            tion. In *Disease in Populations in Transition,* edited by
            Alan C. Swedlund and George J. Armelagos, pp. 127–44.
            Bergin and Garvey, New York.

Armelagos, George J., Alan H. Goodman, and Kenneth H. Jacobs
1991   The Origins of Agriculture: Population Growth During
            a Period of Declining Health. *Population and Environment*
            13:9–22.

Arnold, Jeanne E.
1992   Complex Hunter-Gatherer-Fishers of Prehistoric Califor-
            nia: Chiefs, Specialists, and Maritime Adaptations of the
            Channel Islands. *American Antiquity* 57:60–84.

Asch, David
1994   Prehistoric Plant Husbandry in West-Central Illinois: An
            8000-Year Perspective. In *Agricultural Origins and Develop-
            ment in the Midcontinent,* edited by William Green, pp. 25–
            86. Report No. 19. Office of the State Archaeologist. Uni-
            versity of Iowa, Iowa City.

Asch, David, and Nancy Asch
1977   Chenopod as Cultigen: A Reevaluation of Some Prehis-
            toric Collections from Eastern North America. *Midconti-
            nental Journal of Archaeology* 2:3–45.
1979   Archaeobotany of the Koster Site: The Early and Middle
            Archaic Occupations. Paper presented at the Forty-fourth

Annual Meeting of the Society for American Archaeology. Vancouver, British Columbia.

1985a   Archeobotany of the Campbell Hollow Archaic Occupations. In *The Campbell Hollow Archaic Occupations: A Study of Intrasite Spatial Structure in the Lower Illinois Valley,* edited by C. Russell Stafford, pp. 82–107. Center for American Archaeology, Kampsville Archaeological Center, Kampsville, Illinois.

1985b   Prehistoric Plant Cultivation in West-Central Illinois. In *Prehistoric Food Production in North America,* edited by R. I. Ford, pp. 149–203. Anthropological Papers No. 75. Museum of Anthropology, University of Michigan, Ann Arbor.

1992   Archaeobotany. In *Early Woodland Occupations at the Ambrose Flick Site in the Sny Bottom of West-Central Illinois,* edited by Russell Stafford, pp. 177–293. Research Series No. 10. Kampsville Archeological Center, Kampsville, Illinois.

Asch, Nancy, and David L. Asch

1978   The Economic Potential of *Iva annua* and its Prehistoric Importance in the Lower Illinois Valley. In *The Nature and Status of Ethnobotany,* edited by R. I. Ford, pp. 300–341. Anthropological Papers No. 67. Museum of Anthropology, University of Michigan, Ann Arbor.

Bailey, L. H.

1937   *The Garden of Gourds.* Macmillan Company, New York.

Barbour, Philip L., editor

1986   *The Complete Works of Captain John Smith (1580–1631).* University of North Carolina Press, Chapel Hill.

Barrett, Spencer C. H.

1983   Crop Mimicry in Weeds. *Economic Botany* 37:255–82.

1987   Mimicry in Plants. *Scientific American* 257:76–83.

Bartram, William

1791   *Travels Through North and South Carolina, Georgia, East and West Florida, the Cherokee Country, the Extensive Territories of the Muscogulges, or Creek Confederacy, and the Country of the Chactaws.* James & Johnson, Philadelphia.

1853   Observations on the Creek and Cherokee Indians. Edited by Ephraim G. Squier. *American Ethnological Society, Transactions* III(1):1–81.

1973   *Travels Through North and South Carolina, Georgia, and East and West Florida* [orig 1792]. Bee Hive Press, Savannah, Georgia.

References Cited

Bar-Yosef, Ofer, and Anna Belfer-Cohen
> 1989   The Origins of Sedentism and Farming Communities in the Levant. *Journal of World Prehistory* 3:447–98.
> 1992   From Foraging to Farming in the Mediterranean Levant. In *Transitions to Agriculture in Prehistory,* edited by A. Gebauer and T. Price, pp. 21–48. Prehistory Press, Madison, Wisconsin.

Basgall, Mark E.
> 1987   Resource Intensification Among Hunter-Gatherers: Acorn Economies in Prehistoric California. *Research in Economic Anthropology* 9:21–52.

Battle, Herbert B.
> 1922   The Domestic Use of Oil Among the Southern Aborigines. *American Anthropologist* 24:171–82.

Bender, B.
> 1985   Emergent Tribal Formations in the American Midcontinent. *American Antiquity* 50:52–62.

Berry, Michael
> 1972   *The Evans Site.* A Special Report, Department of Anthropology, University of Utah, Salt Lake City.
> 1985   The Age of Maize in the Greater Southwest: A Critical Review. In *Prehistoric Food Production in North America,* edited by Richard I. Ford, pp. 297–307. Anthropological Papers No. 75. Museum of Anthropology, University of Michigan, Ann Arbor.

Binford, Lewis R.
> 1980   Willow Smoke and Dogs' Tails: Hunter-Gatherer Settlement Systems and Archaeological Site Formation. *American Antiquity* 45:4–20.

Black, Meredith
> 1963   The Distribution and Archaeological Significance of the Marsh Elder *Iva annua* L. *Papers of the Michigan Academy of Science Arts and Letters* XLVIII:541–47.

Bland, E.
> 1651   *The Discovery of New Brittaine.* Thomas Harper, London. Reprinted by Readex Microprint Corporation, 1966.

Bleed, P., Carl Falk, Ann Bleed, and Akira Matsui
> 1989   Between the Mountains and the Sea: Optimal Hunting Patterns and Faunal Remains at Yagi, an Early Jomon Community in Southwestern Hokkaido. *Arctic Anthropology* 26:107–26.

Blurton Jones, Nicholas G.
1990   Three Sensible Paradigms for Research on Evolution and Human Behavior? *Ethnology and Sociobiology* 11:353–59.

Bohrer, Vorsila L.
1962   Nature and Interpretation of Ethnobotanical Materials from Tonto National Monument. In *Archaeological Studies at Tonto National Monument, Arizona,* edited by L. R. Caywood, pp. 75–114. Southwestern Monuments Association Technical Series 2. Globe, Arizona.

1970   Ethnobotanical Aspects of Snaketown, A Hohokam Village in Southern Arizona. *American Antiquity* 35:413–30.

1983   New Life from Ashes: The Tale of the Burnt Bush (*Rhus trilobata*). *Desert Plants* 5:122–24.

1986   Guideposts in Ethnobotany. *Journal of Ethnobiology* 6:27–43.

1987   The Plant Remains from La Ciudad, a Hohokam Site in Phoenix. In *La Ciudad: Specialized Studies in the Economy, Environment, and Culture of La Ciudad,* edited by J. E. Kisselburg, G. E. Rice, and B. L. Shears, pp. 67–200. Arizona State University Office of Cultural Resource Management Anthropological Field Studies No. 20. Tempe, Arizona.

Bozeman, Tandy K.
1982   *Moundville Phase Communities in the Black Warrior River Valley, Alabama.* Ph.D. dissertation, University of California-Santa Barbara. University Microfilms, Ann Arbor, Michigan.

Braidwood, Linda, Robert J. Braidwood, Bruce Howe,
Charles Reed, and Patty Jo Watson, editors
1983   *Prehistoric Archeology Along the Zagros Flanks.* Oriental Institute Publication 105. Oriental Institute, University of Chicago.

Braidwood, Robert J., and Bruce Howe, editors
1960   *Prehistoric Investigations in Iraqi Kurdistan.* Studies in Ancient Oriental Civilization 31. Oriental Institute, University of Chicago.

Braun, E. Lucy
1950   *Deciduous Forests of Eastern North America.* Blakiston, Philadelphia.

Britton, N. L., and A. Brown
1896   *An Illustrated Flora of the Northern United States and Canada and the British Possessions.* Charles Scribner's Sons, New York.

Brown, James
    1985   Long Term Trends to Sedentism and the Emergence of
                    Complexity in the American Midwest. In *Prehistoric
                    Hunter-Gatherers: The Emergence of Cultural Complexity,* ed-
                    ited by T. Price and J. Brown, pp. 201–34. Academic Press,
                    Orlando, Florida.
Brown, James A., and Robert K. Vierra
    1983   What Happened in the Middle Archaic? Introduction to
                    an Ecological Approach to Koster Site Archaeology. In *Ar-
                    chaic Hunters and Gatherers in the American Midwest,* edited
                    by James L. Phillips and James A. Brown, pp. 165–95. Aca-
                    demic Press, New York.
Bryant, Vaughn M., Jr.
    1974   Pollen Analysis of Prehistoric Human Feces from Mam-
                    moth Cave. In *Archeology of the Mammoth Cave Area,*
                    edited by P. J. Watson, pp. 203–9. Academic Press, New
                    York.
Buikstra, Jane E., Jill Bullington, Douglas K. Charles, Della C. Cook, Susan R.
Frankenberg, Lyle W. Konigsberg, Joseph B. Lambers, and Liang Xue
    1987   Diet, Demography, and the Development of Horticulture.
                    In *Emergent Horticultural Economies on the Eastern Wood-
                    lands,* edited by William F. Keegan, pp. 67–85. Southern
                    Illinois University at Carbondale Occasional Paper No. 7.
                    Center for Archaeological Investigations, Carbondale, Illi-
                    nois.
Caldwell, Joseph
    1958   *Trend and Tradition in the Prehistory of the Eastern United
                    States.* American Anthropological Association Memoir 88.
                    Kraus Reprint Company, Millwood, New York.
Cassidy, Claire Monod
    1984   Skeletal Evidence for Prehistoric Subsistence Adaptation
                    in the Central Ohio River Valley. In *Paleopathology at the
                    Origins of Agriculture,* edited by Mark Nathan Cohen and
                    George J. Armelagos, pp. 307–45. Academic Press, New
                    York.
Chapman, Jefferson, and Gary Crites
    1987   Evidence for Early Maize (*Zea mays*) from the Icehouse
                    Bottom Site, Tennessee. *American Antiquity* 52:352–54.
Chapman, Jefferson, and Andrea Brewer Shea
    1981   The Archaeobotanical Record: Early Archaic Period to
                    Contact in the Lower Little Tennessee River Valley. *Ten-
                    nessee Anthropologist* 6:61–84.

Chapman, Jefferson, and Patty Jo Watson

1993   The Archaic Period and the Flotation Revolution. In *Foraging and Farming in the Eastern Woodlands,* edited by C. Margaret Scarry, pp. 27–38. University Press of Florida, Gainesville.

Chavero, E. L., and M. E. Alverez-Buylla Roces

1988   Ethnobotany in a Tropical-Human Region: The Home Gardens of Balzapote, Veracruz, Mexico. *Journal of Ethnobiology* 8:45–79.

Chomko, Stephen A., and Gary W. Crawford

1978   Plant Husbandry in Prehistoric Eastern North America: New Evidence for its Development. *American Antiquity* 43:405–8.

Claggett, Stephen R., and John S. Cable

1982   *The Haw River Sites: Archaeological Investigations at Two Stratified Sites in the North Carolina Piedmont.* Report No. 2386. Commonwealth Associates, Raleigh, North Carolina.

Cleveland, William S.

1985   *The Elements of Graphing Data.* Wadsworth, Monterey, California.

Cohen, Mark Nathan, and George J. Armelagos

1984   Paleopathology at the Origins of Agriculture: Editors' Summation. In *Paleopathology at the Origins of Agriculture,* edited by Mark Nathan Cohen and George J. Armelagos, pp. 585–601. Academic Press, Orlando, Florida.

Coker, William S., and Thomas D. Watson

1986   *Indian Traders of the Southeastern Spanish Borderlands: Panton, Leslie & Company and John Forbes & Company, 1783–1847.* University of West Florida Press, Pensacola.

Colwell, Robert K.

1974   Predictability, Constancy, and Contingency of Periodic Phenomena. *Ecology* 55:1148–53.

Conard, Nicholas, David L. Asch, Nancy B. Asch, David Elmore, Harry Gove, Meyer Rubin, James A. Brown, Michael D. Wiant, Kenneth B. Farnsworth, and Thomas G. Cook

1984   Prehistoric Horticulture in Illinois: Accelerator Radiocarbon Dating of the Evidence. *Nature* 308:443–46.

Cook, Della Collins

1984   Subsistence and Health in the Lower Illinois Valley: Osteological Evidence. In *Paleopathology at the Origins of Agriculture,* edited by Mark Nathan Cohen and George J. Armelagos, pp. 235–69. Academic Press, New York.

Cornett, James W.

1987   Indians and the Desert Fan Palm. *Masterkey: Anthropology of the Americas* 60:12–17.

Correll, Donovan Stewart, and M. C. Johnston

1970   *Manual of the Vascular Plants of Texas.* Texas Research Foundation, Renner.

Cowan, C. Wesley

1975   *An Archaeological Survey and Assessment of the Proposed Red River Reservoir in Powell, Wolfe, and Menifee Counties, Kentucky.* Report submitted to National Park Service, Interagency Archaeological Services, Atlanta, by the University of Kentucky Museum of Anthropology.

1978a   Prehistoric Use and Distribution of Maygrass in Eastern North America: Cultural and Phytogeographical Implications. In *The Nature and Status of Ethnobotany,* edited by R. I. Ford, pp. 263–88. Anthropological Papers No. 67. Museum of Anthropology, University of Michigan, Ann Arbor.

1978b   Seasonal Nutritional Stress in a Late Woodland Population: Suggestions from Some Eastern Kentucky Coprolites. *Tennessee Anthropologist* 3:117–28.

1979a   Excavations at the Haystack Rock Shelters, Powell County, Kentucky. *Midcontinental Journal of Archaeology* 4:1–33.

1979b   *Prehistoric Plant Utilization at the Rogers Rockshelter, Powell County, Kentucky.* Unpublished Master's thesis, Department of Anthropology, University of Kentucky, Lexington.

1985a   *From Foraging to Incipient Food Production: Subsistence Change and Continuity on the Cumberland Plateau of Eastern Kentucky.* Unpublished Ph.D. dissertation, Department of Anthropology, University of Michigan, Ann Arbor.

1985b   Understanding the Evolution of Plant Husbandry in Eastern North America: Lessons from Botany, Ethnography and Archaeology. In *Prehistoric Food Production in North America,* edited by R. I. Ford, pp. 205–43. Anthropological Papers No. 75. Museum of Anthropology, University of Michigan, Ann Arbor.

Cowan, C. Wesley, H. Edwin Jackson, K. Moore, A. Nickelhoff, and T. Smart

1981   The Cloudsplitter Rockshelter, Menifee County, Kentucky: A Preliminary Report. *Southeastern Archaeological Conference Bulletin* 24:60–75.

Cowan, C. Wesley, and Bruce D. Smith
   1993   New Perspectives on a Wild Gourd in Eastern North America. *Journal of Ethnobiology* 13:17–54.

Cowan, C. W., B. D. Smith, and M. P. Hoffman
   1991   New Perspectives on a Free Ranging Cucurbit in Eastern North America. Paper presented at the 14th Annual Meeting of the Society of Ethnobiology, St. Louis.

Cowan, C. W., and P. J. Watson, editors
   1992   *The Origins of Agriculture: An International Perspective.* Smithsonian Institution Press, Washington, D.C.

Crane, H. R.
   1956   University of Michigan Radiocarbon Dates I. *Science* 124:664–72.

Crawford, Gary W.
   1983   *Paleoethnobotany of the Kameda Peninsula Jomon.* Anthropological Papers No. 73. Museum of Anthropology, University of Michigan, Ann Arbor.

   1992   The Transitions to Agriculture in Japan. In *Transitions to Agriculture in Prehistory,* edited by Anne Birgitte Gebauer and T. Douglas Price, pp. 117–32. Prehistory Press, Madison, Wisconsin.

   Forthcoming. Evidence for Jomon Use and Domestication of Barnyard Millet (*Echinochloa utilis*).

Crawford, Gary W., and H. Takamiya
   1992   The Origins and Implications of Late Prehistoric Plant Husbandry in Northern Japan. *Antiquity* 64:889–911.

Crawford, Gary W., and M. Yoshizaki
   1987   Ainu Ancestors and Early Asian Agriculture. *Journal of Archaeological Science* 14:201–13.

Crites, G. D.
   1978   *Paleoethnobotany of the Normandy Reservoir in the Upper Duck River Valley, Tennessee.* Unpublished Master's thesis, Department of Anthropology, University of Tennessee, Knoxville.

   1985   *Middle Woodland Paleoethnobotany of the Eastern Highland Rim of Tennessee: An Evolutionary Perspective on Change in Human-Plant Interaction.* Ph.D. dissertation, Department of Anthropology, University of Tennessee, Knoxville.

   1987   Human-Plant Mutualism and Niche Expression in the Paleoethnobotanical Record: A Middle Woodland Example. *American Antiquity* 52:725–40.

1991 Investigations into Early Plant Domesticates and Food Production in Middle Tennessee: A Status Report. *Tennessee Anthropologist* 16:69–87.

1993 Domesticated Sunflower in Fifth Millennium B.P. Temporal Context: New Evidence from Middle Tennessee. *American Antiquity* 58:146–48.

Curren, Caleb

1984 *The Protohistoric Period in Central Alabama.* Alabama Tombigbee Regional Commission, Camden, Alabama.

D'Andrea, A. C.

1992 *Paleoethnobotany of Later Jomon and Early Yayoi Cultures in Northeastern Japan: Northeastern Aomori and Southwestern Hokkaido.* Unpublished Ph.D. dissertation, Department of Anthropology, University of Toronto, Toronto.

D'Andrea, A. C., G. W. Crawford, M. Yoshizaki, and T. Kudo

1995 Late Jomon Cultigens in Northeastern Japan. *Antiquity* 69:146–52.

Davis, Anthony

1979 Clearance and Agriculture and its Influence on Regional and Local Pollen Rains in the Jomon and Yayoi Periods with Particular Reference to Northern Honshu and Southwestern Hokkaido. Paper presented at the 44th Annual Meeting of the Society for American Archaeology, Vancouver.

Davis, H. A.

1969 A Brief History of Archeological Work in Arkansas up to 1967. *The Arkansas Archeologist* 10:2–4.

Deam, Charles C.

1921 *Trees of Indiana.* Indiana Department of Conservation, Indianapolis.

Dean, Jeffrey S., Robert C. Euler, George J. Gumerman, Fred Plog, Richard H. Hevley, and Thor N. V. Karlstrom

1985 Human Behavior, Demography, and Paleoenvironment on the Colorado Plateaus. *American Antiquity* 50:537–54.

Decker, D. S.

1988 Origin(s), Evolution, and Systematics of *Cucurbita pepo* (Cucurbitaceae). *Economic Botany* 42:3–15.

Decker, D. S., and H. G. Wilson

1986 Numerical Analysis of Seed Morphology in *Cucurbita pepo. Systematic Botany* 11:595–607.

Decker-Walters, Deena
1990 Evidence for Multiple Domestications of *Cucurbita pepo*. In *Biology and Utilization of the Cucurbitaceae,* edited by David M. Bates, Richard W. Robinson, and Charles Jeffrey, pp. 96–101. Cornell University Press, Ithaca, New York.

Decker-Walters, D. S., T. W. Walters, C. W. Cowan, and B. D. Smith
1993 Isozymic Characterization of Wild Populations of *Cucurbita pepo. Journal of Ethnobiology* 13:55–72.

Delcourt, H. R.
1987 The Impact of Prehistoric Agriculture and Land Occupation on Natural Vegetation. *Trends in Ecology and Evolution* 2:39–44.

Delcourt, P., H. Delcourt, P. Cridlebaugh, and J. Chapman
1986 Holocene Ethnobotanical and Paleoecological Record of Human Impact on Vegetation in the Little Tennessee River Valley, Tennessee. *Quaternary Research* 25:330–49.

Densmore, Frances
1928 *Uses of Plants by the Chippewa Indians.* Forty-Fourth Annual Report of the Bureau of American Ethnology to the Secretary of the Smithsonian Institution, 1926–27. U.S. Government Printing Office, Washington, D.C.

Dickinson, Martin F., and Lucy B. Wayne
1985 The Seminole Indian Dispersed Settlement Pattern: An Example from Marion County, Florida. In *Indians, Colonists, and Slaves: Essays in Memory of Charles H. Fairbanks,* edited by K. W. Johnson, J. M. Leader, and R. C. Wilson, pp. 221–39. Florida Anthropology Student Association, Gainesville.

Dodd, Walter A., Jr.
1982 *Final Year Excavations at the Evans Mound Site.* University of Utah Anthropological Papers No. 106. Salt Lake City.

Doolittle, William E.
1992 Agriculture in North America on the Eve of Contact: A Reassessment. In "The Americas Before and After 1492: Current Geographical Research," edited by Karl W. Butzer. *Annals of the Association of American Geographers* 82:386–401.

Doran, Glen H., David N. Dickel, and Lee A. Newsom
1990 A 7,290-Year-Old Bottle Gourd from the Windover Site, Florida. *American Antiquity* 55:354–59.

Doran, James E., and F. Roy Hodson
    1975   *Mathematics and Computers in Archaeology.* Harvard University Press, Cambridge.

Douglas, Brian J.
    1985   The Biology of Canadian Weeds; *Setaria viridis* (L.) Beauv. *Canadian Journal of Plant Science* 65:669–90.

Downs, Albert A., and William E. McQuilkin
    1944   Seed Production in Southern Appalachian Oaks. *Journal of Forestry* 42:913–20.

Driver, Harold E.
    1969   *Indians of North America.* 2nd ed. University of Chicago Press, Chicago.

Dunnell, Robert C.
    1980   Evolutionary Theory and Archaeology. In *Advances in Archaeological Method and Theory,* vol. 3, edited by Michael B. Schiffer, pp. 35–99. Academic Press, New York.

Durham, William H.
    1992   Applications of Evolutionary Culture Theory. *Annual Review of Anthropology* 21:331–55.

Elster, Jon
    1983   *Explaining Technical Change.* Cambridge University Press, Cambridge.
    1985   *Making Sense of Marx.* Cambridge University Press, Cambridge.

Emerson, Thomas E.
    1992   The Mississippian Dispersed Village as a Social and Environmental Strategy. In *Late Prehistoric Agriculture: Observations from the Midwest,* edited by W. I. Woods, pp. 1–18. Studies in Illinois Archaeology No. 8. Illinois Historic Preservation Agency, Springfield, Illinois.

Ensor, H. Blaine
    1993   *Big Sandy Farms: A Prehistoric Agricultural Community Near Moundville, Black Warrior Valley Floodplain, Tuscaloosa County, Alabama.* Report of Investigations No. 68. Alabama Museum of Natural History Division of Archaeology, Tuscaloosa.

Fiedel, S. J.
    1987   *Prehistory of the Americas.* Cambridge University Press, London.

Fish, Paul R.
    1989   The Hohokam: 1,000 Years of Prehistory in the Sonoran Desert. In *Dynamics of Southwest Prehistory,* edited by

L. Cordell and G. Gumerman, pp. 19–63. Smithsonian Institution Press, Washington, D.C.

Flannery, Kent V.

1965   The Ecology of Early Food Production in Mesopotamia.
        Science 147:1247–56.

1969   Origins and Ecological Effects of Early Domestication
        in Iran and the Near East. In *The Domestication and Exploi-
        tation of Plants and Animals,* edited by Peter J. Ucko and
        George W. Dimbleby, pp. 73–100. Aldine, Chicago.

1971   Archaeological Systems Theory and Early Mesoamerica.
        In *Prehistoric Agriculture,* edited by Stuart Struever,
        pp. 80–100. Natural History Press, Garden City, New
        Jersey.

1986   Adaptation, Evolution, and Archaeological Phases: Some
        Implications of Reynolds' Simulation. In *Guilá Naquitz,*
        edited by Kent Flannery, pp. 501–7. Academic Press, Or-
        lando, Florida.

Flannery, Kent V., editor

1986   *Guilá Naquitz: Archaic Foraging and Early Agriculture in
        Oaxaca, Mexico.* Academic Press, Orlando, Florida.

Food and Agricultural Organization of the United Nations

1974   *Handbook of Human Nutritional Requirements.* FAO Nutri-
        tional Studies No. 228. Rome, Italy.

Ford, Richard I.

1977   Evolutionary Ecology and the Evolution of Human
        Ecosystems: A Case Study from the Midwestern U.S.A.
        In *Explanation of Prehistoric Change,* edited by James N.
        Hill, pp. 153–84. University of New Mexico Press, Albu-
        querque.

1979   Paleoethnobotany in American Archaeology. *Advances in
        Archaeological Method and Theory* 2:285–336.

1981   Gardening and Farming Before A.D. 1000: Patterns of Pre-
        historic Cultivation North of Mexico. *Journal of Ethnobiol-
        ogy* 1:6–27.

1985   Patterns of Food Production in Prehistoric North Amer-
        ica. In *Prehistoric Food Production in North America,* edited
        by R. Ford, pp. 341–64. Anthropological Papers No. 75.
        Museum of Anthropology, University of Michigan, Ann
        Arbor.

1986   Reanalysis of Cucurbits in the Ethnobotanical Labora-
        tory, University of Michigan. *The Missouri Archaeologist*
        47:13–31.

1987   Dating Early Maize in the Eastern United States. Paper presented at the 10th Ethnobiology Conference, Gainesville, Florida.

Ford, Richard I., editor

1985   *Prehistoric Food Production in North America.* Anthropological Papers No. 75. Museum of Anthropology, University of Michigan, Ann Arbor.

Forman, R. T. T., and M. Godron

1986   *Landscape Ecology.* John Wiley and Sons, New York.

Fowells, H. A.

1965   *Silvics of Forest Trees of the United States.* Agricultural Handbook No. 271. U.S. Department of Agriculture, Washington, D.C.

Fowler, Catherine

1986   Subsistence. In *Handbook of North American Indians.* Vol. 11 of *Great Basin,* edited by Warren L. D'Azevedo, pp. 64–97. Smithsonian Institution, Washington, D.C.

1993   *In the Shadow of Fox Peak: An Ethnography of the Cattail-Eater Northern Paiute People of Stillwater Marsh.* Cultural Resource Series No. 5. U.S. Department of the Interior, Fish and Wildlife Service, Region 1, Stillwater National Wildlife Refuge, U.S. Government Printing Office, Washington, D.C.

1994   Historical Perspectives on Timbisha Shoshone Land Management Practices, Death Valley, CA. Paper presented at the Annual Meeting of the Society of Ethnobiology, Victoria, British Columbia.

Fowler, Melvin L.

1969   Middle Mississippian Agricultural Fields. *American Antiquity* 34:365–75.

1992   The Eastern Horticultural Complex and Mississippian Agricultural Fields: Studies and Hypotheses. In *Late Prehistoric Agriculture: Observations from the Midwest,* edited by W. I. Woods, pp. 1–18. Studies in Illinois Archaeology No. 8. Illinois Historic Preservation Agency, Springfield, Illinois.

Fritz, G. J.

1986   *Prehistoric Ozark Agriculture: The University of Arkansas Rockshelter Collections.* Ph.D. dissertation, University of North Carolina at Chapel Hill. University Microfilms, Ann Arbor, Michigan.

1990   Multiple Pathways to Farming in Precontact Eastern North America. *Journal of World Prehistory* 4:387–435.

1994   The Value of Archeological Plant Remains for Paleodietary Reconstruction. In *Paleonutrition: The Diet and Health of Prehistoric Americans,* edited by K. D. Sobolik, pp. 21–33. Occasional Paper No. 22. Center for Archaeological Investigations, Southern Illinois University at Carbondale.

Fritz, Gayle J., and Bruce D. Smith

1988   Old Collections and New Technology: Documenting the Domestication of *Chenopodium* in Eastern North America. *Midcontinental Journal of Archaeology* 13:3–27.

Funkhouser, William D., and William S. Webb

1929   *The So-called "Ash Caves" in Lee County, Kentucky.* University of Kentucky Reports in Archaeology and Anthropology 1(2). Lexington.

1930   *Rock Shelters of Wolfe and Powell Counties, Kentucky.* University of Kentucky Reports in Archaeology and Anthropology 1(4). Lexington.

Gardner, Paul S.

1992   *Diet Optimization Models and Prehistoric Subsistence Change in the Eastern Woodlands.* Unpublished Ph.D. dissertation, Department of Anthropology, University of North Carolina, Chapel Hill.

Gardner, Peter M.

1991   Forager's Pursuit of Individual Autonomy. *Current Anthropology* 32:543–72.

Gebauer, A. B., and D. Price, editors

1992   *Transitions to Agriculture in Prehistory.* Prehistory Press, Madison, Wisconsin.

Gilmore, Melvin R.

1931   Vegetal Remains of the Ozark Bluff-Dweller Culture. *Papers of the Michigan Academy of Science, Arts, and Letters* 14:83–102.

1977   *Uses of Plants by the Indians of the Missouri River Region* [orig. 1919]. University of Nebraska Press, Lincoln.

Goland, Carol

1991   The Ecological Context of Hunter-Gatherer Storage: Environmental Predictability and Environmental Risk. In *Foragers in Context,* edited by Preston T. Miracle, Lynn E. Fisher, and Jody Brown, pp. 107–25. Michigan Discussions in Anthropology No. 10. Ann Arbor.

1992a   Cultivating Diversity: Field Scattering as Agricultural Risk Management in Cuyo Cuyo, Dept. of Puno, Peru. Working Paper No. 4. PSE Project, Department of Anthropology, University of North Carolina, Chapel Hill.

1992b   Field Scattering as Agricultural Risk Management: A
             Case Study from Cuyo Cuyo, Department of Puno, Peru.
             MS. on file.

Goldschmidt, Walter
1974   Subsistence Activities Among the Hupa. In *Indian Land
             Use and Occupancy in California,* Volume 1, edited by
             R. Beals and J. Hester, Jr., pp. 52–55. Garland Publishing,
             New York.

Goodman, Alan H., and George J. Armelagos
1985   Disease and Death at Dr. Dickson's Mounds. *Natural His-
             tory* 94:12–18.

Goodman, Alan H., John Lallo,
George J. Armelagos, and Jerome C. Rose
1984   Health Changes at Dickson Mounds, Illinois (A.D. 950–
             1300). In *Paleopathology at the Origins of Agriculture,* edited
             by Mark Nathan Cohen and George J. Armelagos, pp.
             271–305. Academic Press, New York.

Goodrum, P. D., V. H. Reid, and C. E. Boyd
1971   Acorn Yields, Characteristics, and Management Criteria of
             Oaks for Wildlife. *Journal of Wildlife Management* 35:520–32.

Gowlett, J. A. J.
1987   The Archaeology of Radiocarbon Accelerator Dating.
             Journal of World Prehistory 1:127–70.

Grant, Hugh, editor
1980   *Letters, Journals and Writings of Benjamin Hawkins.* 2 vols.
             Beehive Press, Savannah, Georgia.

Gremillion, Kristen J.
1993a   Crop and Weed in Prehistoric Eastern North America:
             The *Chenopodium* Example. *American Antiquity* 58:496–508.
1993b   Paleoethnobotany. In *The Development of Southeastern Ar-
             chaeology,* edited by Jay Johnson, pp. 132–59. University of
             Alabama Press, Tuscaloosa.
1993c   Plant Husbandry at the Archaic/Woodland Transition:
             Evidence from the Cold Oak Shelter, Kentucky. *Midconti-
             nental Journal of Archaeology* 18:161–89.
1994   Botanical Contents of Paleofeces from Three Eastern
             Kentucky Rockshelters. Paper presented at the Annual
             Meeting of the Kentucky Heritage Council, Lexington.

Gunn, Joel
1991   Influences of Various Forcing Variables on Global Energy
             Balance During the Period of Intensive Instrumental Ob-
             servation (1958–1987) and Their Implications for Paleocli-
             mate. *Climatic Change* 19:393–420.

Haag, W. G.
1974   The Adena Culture. In *Archaeological Researches in Retrospect,* edited by G. R. Willey, pp. 119–45. Winthrop Publishers, Cambridge.
Hajic, Edwin R.
1981   *Geology and Paleopedology of the Koster Archaeological Site, Greene County, Illinois.* Unpublished Master's thesis, Department of Geology, University of Iowa, Iowa City.
Hall, Robert L.
1977   An Anthropocentric Perspective for Eastern United States Prehistory. *American Antiquity* 42:499–518.
Halstead, Paul, and John O'Shea
1989   Introduction: Cultural Responses to Risk and Uncertainty. In *Bad Year Economics,* edited by Paul Halstead and John O'Shea, pp. 1–7. Cambridge University Press, Cambridge.
Hammett, J. E.
1990   An Analysis of Plant Remains from 1987 Excavations at Talepop (Ca-LAn-229). In *Archaeological Studies at Site CA-LAn-229, Malibu Creek State Park: An Experiment in Inference Justification,* edited by Mark Raab, California Northridge Center for Public Archaeology, California State University at Northridge, for California State Department of Parks and Recreation, Interagency Agreement No. 40-39-012.
1991   *The Ecology of Sedentary Societies Without Agriculture: Paleoethnobotanical Indicators from Native California.* Ph.D. dissertation, Department of Anthropology, University of North Carolina, Chapel Hill.
1992   Ethnohistory of Aboriginal Landscapes and Land Use in the Southeastern United States. *Southern Indian Studies* 41:1–50.
Harlan, Jack R.
1975   *Crops and Man.* American Society of Agronomy, Madison, Wisconsin.
Harper, Francis, editor
1958   *The Travels of William Bartram.* Naturalist's edition. Yale University, New Haven.
Harper, Roland H.
1944   *Preliminary Report on the Weeds of Alabama.* Bulletin 53. Geological Survey of Alabama, University of Alabama, University.

Harris, D., and Gordon Hillman, editors
  1989   *Foraging and Farming: The Evolution of Plant Exploitation.*
                Allen and Unwin, London.
Hastorf, C. A.
  1988   The Use of Paleoethnobotanical Data in Prehistoric
                Studies of Crop Production, Processing, and Consump-
                tion. In *Current Paleoethnobotany: Analytical Methods and
                Cultural Interpretations of Archaeological Plant Remains,* ed-
                ited by C. A. Hastorf and V. S. Popper, pp. 119–44. Univer-
                sity of Chicago Press, Chicago.
  1991   Gender, Space, and Food in Prehistory. In *Engendering
                Archaeology: Women and Prehistory,* edited by J. Gero and
                M. Conkey, pp. 132–59. Basil Blackwell, Oxford.
Hatley, M. Thomas
  1989   The Three Lives of Keowee: Loss and Recovery in Eigh-
                teenth-Century Cherokee Villages. In *Powhatan's Mantle:
                Indians in the Colonial Southeast,* edited by P. H. Wood,
                G. A. Waselkov, and M. T. Hatley, pp. 223–48. University
                of Nebraska Press, Lincoln.
  1993   *The Dividing Paths: Cherokees and South Carolinians
                through the Era of Revolution.* Oxford University Press,
                New York.
Hawkes, K., and J. O'Connell
  1992   On Optimal Foraging Models and Subsistence Transi-
                tions. *Current Anthropology* 33:63–66.
Heiser, C. B., Jr.
  1954   Variation and Subspeciation in the Common Sunflower,
                Helianthus annuus. American Midland Naturalist 51:287–305.
  1969   The North American Sunflowers (*Helianthus*). *Memoirs of
                the Torrey Botanical Club* 22:1–217.
  1976   *The Sunflower.* University of Oklahoma Press, Norman.
  1978   Taxonomy of *Helianthus* and Origin of Domesticated
                Sunflower. In *Sunflower Science and Technology,* edited by
                J. F. Carter, pp. 31–53. American Society of Agronomy
                No. 19. Madison, Wisconsin.
  1985   Some Botanical Considerations of Early Domesticated
                Plants North of Mexico. In *Prehistoric Food Production in
                North America,* edited by R. I. Ford, pp. 57–72. Anthropo-
                logical Papers No. 75. Museum of Anthropology, Univer-
                sity of Michigan, Ann Arbor.
  1989   Domestication of Cucurbitaceae: *Cucurbita* and *Lagenaria.*
                In *Foraging and Farming: The Evolution of Plant Exploita-*

tion, edited by D. R. Harris and G. C. Hillman, pp. 470–80. Unwin Hyman, London.

Helbaek, Hans

1960   The Paleoethnobotany of the Near East and Europe. In *Prehistoric Investigations in Iraqi Kurdistan,* edited by R. Braidwood and B. Howe, pp. 99–118. Studies in Ancient Oriental Civilization No. 31. Oriental Institute, University of Chicago.

1963   Palaeo-Ethnobotany. In *Science in Archaeology,* edited by D. Brothwell and E. Higgs, pp. 177–85. Basic Books, New York.

1966   Pre-Pottery Neolithic Farming at Beidha. In "Five Seasons at the Pre-Pottery Neolithic Village at Beidha in Jordan," by Diana Kirkbride. *Palestine Exploration Quarterly* 98:8–72.

1969   Plant Collecting, Dry-Farming and Irrigation Agriculture in Prehistoric Deh Luran. In *Prehistory and Human Ecology of the Deh Luran Plain: An Early Village Sequence from Khuzistan,* edited by F. Hole, K. Flannery, and J. Neely, pp. 383–426. Memoir No. 1. Museum of Anthropology, University of Michigan, Ann Arbor.

Henbest, Wayne

1934   Unpublished Field Notes of the Excavations at Marble Bluff Rockshelter. *University of Arkansas Museum Field Book* 13:55–109.

Henri, Florette

1986   *The Southern Indians and Benjamin Hawkins, 1796–1816.* University of Oklahoma Press, Norman.

Henry, Donald O.

1989   *From Foraging to Agriculture: The Levant at the End of the Ice Age.* University of Pennsylvania Press, Philadelphia.

Hickerson, H.

1965   The Virginia Deer and Inter-Tribal Buffer Zones in the Upper Mississippi Valley. In *Man, Culture and Animals: The Role of Animals in Human Ecological Adjustments,* edited by A. Leeds and A. P. Vayda, pp. 43–65. American Association for the Advancement of Sciences Publication No. 78. Washington, D.C.

Hill, Kim

1988   Macronutrient Modification of Optimal Foraging Theory: An Approach Using Indifference Curves Applied to Some Modern Foragers. *Human Ecology* 16:157–97.

Hillman, Gordon C., and M. Stuart Davies
    1990   Measured Domestication Rates in Wild Wheats and
                Barley under Primitive Cultivation, and their Archae-
                ological Implications. *Journal of World Prehistory* 4:157–
                222.

Hitchcock, A. S.
    1971   *Manual of the Grasses of the United States.* 2d ed., Revised
                by Agnes Chase, 2 vols. Dover, New York.

Hitchcock, C. L., and A. Cronquist
    1973   *Flora of the Pacific Northwest.* University of Washington
                Press, Seattle.

Hodder, Ian
    1989   *Reading the Past: Current Approaches to Interpretation in
                Archaeology.* 2d ed. Cambridge University Press, Cam-
                bridge.

Hogan, Patrick
    1980   The Analysis of Human Coprolites from Cowboy
                Cave. In *Cowboy Cave,* by Jesse D. Jennings, pp. 201–11.
                University of Utah Anthropological Papers 104. Salt
                Lake City.

Howell, David L.
    1994   Ainu Ethnicity and the Boundaries of the Early Modern
                Japanese State. *Past and Present* 142:69–93.

Hudson, Charles
    1976   *The Southeastern Indians.* University of Tennessee Press,
                Knoxville.

Ison, C. R.
    1988   The Cold Oak Shelter: Providing a Better Understanding
                of the Terminal Archaic. In *Paleoindian and Archaic Re-
                search in Kentucky,* edited by Charles D. Hockensmith,
                David Pollack, and Thomas N. Sanders, pp. 205–20. Ken-
                tucky Heritage Council, Frankfort.
    1991   Prehistoric Upland Farming Along the Cumberland Pla-
                teau. In *Studies in Kentucky Archaeology,* edited by
                Charles D. Hockensmith, pp. 1–10. Kentucky Heritage
                Council, Frankfort.

Jarman, H. N., A. J. Legge, and H. A. Charles
    1972   Retrieval of Plant Remains from Archaeological Sites by
                Froth Flotation. In *Papers in Economic Prehistory,* edited by
                E. Higgs, pp. 39–48. Cambridge University Press, Cam-
                bridge.

Jenkins, Ned J., and Jerry J. Nielsen
  1974   Archaeological Salvage Investigations at the West
                Jefferson Steam Plant Site, Jefferson County, Alabama.
                Ms. on file, Mound State Monument, Moundville, Ala-
                bama.
Jepson, Willis Linn
  1970   *Manual of the Flowering Plants of California.* University of
                California Press, Berkeley.
Johannessen, S.
  1984   Paleoethnobotany. In *American Bottom Archaeology,* edited
                by C. B. Bareis and J. W. Porter, pp. 197–214. University
                of Illinois Press, Urbana.
Johnson, Jay K., Patricia K. Galloway, and Walter Belokon
  1989   Historic Chickasaw Settlement Patterns in Lee County,
                Mississippi: A First Approximation. *Mississippi Archaeology*
                24(2):45–52.
Jones, Volney
  1936   The Vegetal Remains of the Newt Kash Hollow Shelter.
                In *Rock Shelters in Menifee County, Kentucky,* edited by
                W. D. Funkhouser and W. S. Webb, pp. 147–65. Reports in
                Archaeology and Anthropology No. 3(4). University of
                Kentucky, Lexington.
Jones, Walter B., and David L. DeJarnette
  n.d.   *Moundville Culture and Burial Museum.* Geological Survey
                of Alabama Museum Paper 13. University of Alabama,
                University.
Kaplan, Hillard, and Kim Hill
  1985   Food Sharing Among Aché Foragers: Tests of Explanatory
                Hypotheses. *Current Anthropology* 26:223–46.
Kasahara, Y., M. Buda, and A. Fujizawa.
  1986   Yonago-shi Megumi iseki no shushi bunseki dotei
                (Analysis and Identification of Seeds from the Megumi
                Site, Yonago City). In *Megumi iseki,* edited by Yonago-shi
                Kyoiku Iinkai, pp. 98–128. Yonago-shi Kyoiki Iinkai,
                Yonago-shi, Japan.
Kay, M., F. B. King, and C. K. Robinson
  1980   Cucurbits from Phillips Spring: New Evidence and Inter-
                pretations. *American Antiquity* 45:802–22.
Keegan, William F., and Brian M. Butler
  1987   The Microeconomic Logic of Horticultural Intensifica-
                tion in the Eastern Woodlands. In *Emergent Horticultural*

Economies on the Eastern Woodlands, edited by William F. Keegan, pp. 109–27. Occasional Paper No. 7. Center for Archaeological Investigations, Southern Illinois University at Carbondale.

Keene, Arthur S.

1979   *Prehistoric Hunter-Gatherers of the Deciduous Forest: A Linear Programming Approach to Late Archaic Subsistence in the Saginaw Valley (Michigan).* Unpublished Ph.D. dissertation, University of Michigan, Ann Arbor.

1981   Optimal Foraging in a Nonmarginal Environment: A Model of Prehistoric Subsistence Strategies in Michigan. In *Hunter-Gatherer Foraging Strategies: Ethnographic and Archeological Analyses,* edited by Bruce Winterhalder and Eric Alden Smith, pp. 171–93. University of Chicago Press, Chicago.

Kelly, Arthur R.

1938   *A Preliminary Report on Archaeological Explorations at Macon, Georgia.* Bureau of American Ethnology Bulletin 119. Smithsonian Institution, Washington, D.C.

Killion, Thomas W., editor

1992   *Gardens of Prehistory: The Archaeology of Settlement Agriculture in Greater Mesoamerica.* University of Alabama Press, Tuscaloosa.

Kimber, C.

1973   Spatial Patterning in the Dooryard Garden in Puerto Rico. *Geographical Review* 63:6–26.

King, Chester

1990   *Evolution of Chumash Society: A Comparative Study of Artifacts Used for Social System Maintenance in the Santa Barbara Channel Regions Before A.D. 1804.* Garland, New York.

King, Frances B.

1985   Early Cultivated Cucurbits in Eastern North America. In *Prehistoric Food Production in North America,* edited by Richard I. Ford, pp. 73–98. Anthropological Papers No. 75. Museum of Anthropology, University of Michigan, Ann Arbor.

1987   The Evolutionary Effects of Plant Cultivation. In *Emergent Horticultural Economies of the Eastern Woodlands,* edited by William F. Keegan, pp. 51–65. Occasional Paper No. 7. Center for Archaeological Investigations, Southern Illinois University at Carbondale.

References Cited

Knight, Vernon James, Jr.
  1989   Certain Aboriginal Mounds at Moundville: 1937 Excava-
              tions in Mounds H, I, J, K, and L. Paper presented at the
              46th Annual Southeastern Archaeological Conference,
              Tampa.

Knight, Vernon James, Jr., and Carlos Solis
  1983   The Farmstead Papers II: Mississippian Farmsteads and
              Their Economic Significance in the Southeast. Paper pre-
              sented at the 16th Alabama Academy of Sciences, Tus-
              caloosa.

Ladd, Edmund J.
  1979   Zuni Economy. In *Handbook of North American Indians.*
              Vol. 9 of *Southwest,* edited by Alfonso Ortiz, pp. 492–98.
              Smithsonian Institution, Washington, D.C.

Lakela, Olga
  1965   *An Illustrated Flora of Northeast Minnesota.* University of
              Minnesota Press, Minneapolis.

Lankford, George E.
  1983   Ethnohistory: A Documentary Study of Native Ameri-
              can Life in the Lower Tombigbee Valley. In *Cultural Re-
              sources Reconnaissance Study of the Black Warrior–Tombigbee
              System Corridor, Alabama,* vol. 2, edited by Eugene Wilson.
              University of South Alabama, Mobile.

Larsen, Clark Spencer
  1984   Health and Disease in Prehistoric Georgia: The Transition
              to Agriculture. In *Paleopathology at the Origins of Agricul-
              ture,* edited by Mark Nathan Cohen and George J.
              Armelagos, pp. 367–92. Academic Press, New York.

Layton, Robert, and Robert Foley
  1992   On Subsistence Transitions. Response to Hawkes and
              O'Connell. *Current Anthropology* 33:218–19.

Layton, Robert, Robert Foley, and Elizabeth Williams
  1991   The Transition Between Hunting and Gathering and the
              Specialized Husbandry of Resources. *Current Anthropology*
              32:255–74.

Leacock, Eleanor
  1954   *The Montagnais "Hunting Territory" and the Fur Trade.*
              American Anthropological Association Memoir No. 78.
              Menasha, Wisconsin.

Lefler, H. T., editor
  1967   *A New Voyage to Carolina,* by John Lawson [reprint of 1709
              edition]. University of North Carolina Press, Chapel Hill.

Lewis, H. T.
 1973   *Patterns of Indian Burning in California: Ecology and Ethno-history.* Anthropological Papers No. 1. Ballena Press, Ramona, California.
 1977   Maskuta: The Ecology of Indian Fires in Northern Alberta. *The Western Canadian Journal of Anthropology* 70:15–52.

Lewis, H. T., and T. A. Ferguson
 1988   Yards, Corridors and Mosaics: How to Burn a Boreal Forest. *Human Ecology* 16:57–77.

Looman, J., and K. F. Best
 1987   *Budd's Flora of the Canadian Prairie Provinces.* Research Branch Agriculture Canada Publication No. 1662. Hull, Quebec.

Lopinot, Neal H.
 1982   Plant Macroremains and Paleoethnobotanical Implications. In *The Carrier Mills Archaeological Project: Human Adaptation in the Saline Valley, Illinois,* vol. 2, edited by Richard W. Jeffries and Brian M. Butler, pp. 671–855. Center for Archaeological Investigations Research Paper No. 33. Southern Illinois University, Carbondale.

Lynch, Thomas F.
 1973   Harvest Timing, Transhumance, and the Process of Domestication. *American Anthropologist* 75:1254–59.

MacArthur, Robert H., and Eric R. Pianka
 1966   On Optimal Use of Patchy Environments. *The American Naturalist* 100:603–9.

McCloskey, Donald N.
 1975   The Persistence of English Common Fields. In *European Peasants and Their Markets,* edited by William N. Parker and Eric L. Jones, pp. 73–119. Princeton University Press, Princeton, New Jersey.
 1976   English Open Fields as Behavior Towards Risk. In *Research in Economic History,* vol. 1, edited by P. Uselding, pp. 124–70. JAI, Greenwich, Connecticut.

McCorriston, Joy, and Frank Hole
 1991   The Ecology of Seasonal Stress and the Origins of Agriculture in the Near East. *American Anthropologist* 93:39–62.

McKenzie, Douglas H.
 1964   *The Moundville Phase and Its Position in Southeastern Prehistory.* Unpublished Ph.D. dissertation, Department of Anthropology, Harvard University, Cambridge.

1966   Summary of the Moundville Phase. *Journal of Alabama Archaeology* 12:1–58.

McMinn, H. E., and E. Maino
1963   *An Illustrated Manual of Pacific Coast Trees.* University of California Press, Berkeley.

McVickar, Janet L.
1991   *Holocene Vegetation Change at Cowboy Cave, Southeastern Utah and Its Effect on Human Subsistence.* Master's thesis, Quaternary Studies Program, University of Northern Arizona, Flagstaff.

Marquardt, William H.
1974   A Statistical Analysis of Constituents in Human Paleofecal Specimens from Mammoth Cave. In *Archeology of the Mammoth Cave Area,* edited by P. J. Watson, pp. 193–202. Academic Press, New York.

Martin, W. C., and C. R. Hutchins
1981   *A Flora of New Mexico.* 2 vols. A. R. Gantner Verlag, Germany.

Matson, R. G.
1991   *The Origins of Southwestern Agriculture.* University of Arizona Press, Tucson.

Mauelshagen, Carl, and Gerald H. Davis, translators and editors
1969   *Partners in the Lord's Work: The Diary of Two Moravian Missionaries in the Creek Indian Country, 1807–1813.* Research Paper No. 21. Georgia State College, Atlanta.

Maun, M. A., and S. C. H. Barrett
1986   The Biology of Canadian Weeds: *Echinochloa crus-galli* (L.) Beauv. *Canadian Journal of Plant Science* 66:739–59.

Mellars, P.
1976   Fire Ecology, Animal Populations and Man: A Study of Some Ecological Relationships in Prehistory. *Proceedings of the Prehistoric Society* 42:15–45.

Merrill, William L.
1979   The Beloved Tree: *Ilex vomitoria* Among the Indians of the Southeast and Adjacent Regions. In *Black Drink: A Native American Tea,* edited by Charles Hudson, pp. 40–82. University of Georgia Press, Athens.

Michals, Lauren M.
1990   Faunal Exploitation and Chiefdom Organization at Moundville, Alabama. Paper presented at the 55th Annual Meeting of the Society for American Archaeology, Las Vegas.

Milanich, Jerald T., editor

1991   *The Hernando de Soto Expedition*. Spanish Borderlands Sourcebooks No. 11. Garland Publishing, New York.

Miller, Naomi F.

1988   Ratios in Paleoethnobotanical Analysis. In *Current Paleoethnobotany: Analytical Methods and Cultural Interpretations of Archaeological Plant Remains,* edited by C. A. Hastorf and V. S. Popper, pp. 72–85. University of Chicago Press, Chicago.

1991   The Near East. In *Progress in Old World Paleoethnobotany,* edited by W. van Zeist, K. Wasylykowa, and K.-E. Behre, pp. 133–60. A. A. Balkema, Rotterdam.

1992   The Origins of Plant Cultivation in the Near East. In *The Origins of Agriculture: An International Perspective,* edited by C. W. Cowan and P. J. Watson, pp. 39–58. Smithsonian Institution Press, Washington, D.C.

Minnis, Paul E.

1985a   Domesticating Plants and People in the American Southwest. In *Prehistoric Food Production in North America,* edited by R. Ford, pp. 309–40. Anthropological Papers No. 75. Museum of Anthropology, University of Michigan, Ann Arbor.

1985b   *Social Adaptation to Food Stress: A Prehistoric Example.* University of Chicago Press, Chicago.

1992   Early Plant Cultivation in the Desert Borderlands of the American West. In *The Origins of Agriculture: An International Perspective,* edited by W. Cowan and P. Watson, pp. 121–41. Smithsonian Institution Press, Washington, D.C.

Moore, Clarence B.

1905   Certain Aboriginal Remains of the Black Warrior River. *Journal of the Philadelphia Academy of Natural Sciences* 13:123–224.

1907   Moundville Revisited. *Journal of the Philadelphia Academy of Natural Sciences* 13:337–405.

Moratto, Michael J.

1984   The Central Valley Region. In *California Archaeology,* edited by Michael J. Moratto, pp. 167–216. Academic Press, Orlando, Florida.

Morton, Thomas

1637   *New English Canaan or New Canaan.* Jacob Frederick Stam, Amsterdam (reprinted in 1972 by Arno Press, New York).

Muller, Jon, and J. E. Stephens
    1991   Mississippian Sociocultural Adaptation. In *Cahokia and the Hinterlands,* edited by Thomas E. Emerson and R. Barry Lewis, pp. 297–310. University of Illinois Press, Urbana.

Munson, Patrick J.
    1986   Hickory Silviculture: A Subsistence Revolution in the Prehistory of Eastern North America. Paper presented at the Conference on Emergent Horticultural Economies of the Eastern Woodlands, Carbondale, Illinois.

    1988   Late Woodland Settlement and Subsistence in Temporal Perspective. In *Interpretations of Culture Change in the Eastern Woodlands During the Late Woodland Period,* edited by Richard W. Yerkes, pp. 7–16. The Ohio State University Department of Anthropology Occasional Papers in Anthropology No. 3. Columbus.

Munz, Philip A., and David D. Keck
    1973   *A California Flora with Supplement.* University of California Press, Berkeley.

Newsom, Lee
    1988   Paleoethnobotanical Remains from a Waterlogged Archaic Period Site in Florida. Paper presented at the 53rd Annual Meeting of the Society for American Archaeology, Phoenix, Arizona.

Newsom, Lee, S. David Webb, and James S. Dunbar
    1993   History and Geographic Distribution of *Cucurbita pepo* Gourds in Florida. *Journal of Ethnobiology* 13:75–97.

Nishida, Masaki
    1980   Jomon-jidai no shokuryo shigen to seigyo-katsudo: Torihama Kaizuka no shizen ibutsu o chushin to shite (Food Resources and Subsistence Activity During the Jomon Period Based on the Natural Remains from the Torihama Shellmound). *Quaternary Anthropology* II-3:3–56.

    1981   Jomon-jidai no ningen-shokubutsu kankei: shokuryo seisan no shutsugen katei (Man-Plant Relationships in the Jomon Period and the Emergence of Food Production). *Bulletin of the National Museum of Ethnology* 6:234–55.

    1983   The Emergence of Food Production in Neolithic Japan. *Journal of Anthropological Archaeology* 2:305–22.

Nixon, Charles M., Milford W. McClain, and Robert W. Donohoe
    1975   Effects of Hunting and Mast Crops on a Squirrel Population. *Journal of Wildlife Management* 39:1–25.

Odum, Eugene P.
   1971   *Fundamentals of Ecology.* W. B. Saunders, Toronto.
Oetelaar, Gerald
   1990   Faunal Analysis. In *Childers and Woods: Two Late Woodland Sites in the Upper Ohio Valley, Mason County, Kentucky,* edited by Michael J. Shott, pp. 617–90. University of Kentucky Program for Cultural Resource Assessment Archaeological Report No. 200. Lexington.

O'Hear, John W.
   1975   *Site 1Je32: Community Organization in the West Jefferson Phase.* Unpublished Master's thesis, Department of Anthropology, University of Alabama, Tuscaloosa.

Orlove, Benjamin S., and Ricardo Godoy
   1986   Sectoral Fallowing Systems in the Central Andes. *Journal of Ethnobiology* 6:169–204.

O'Shea, J.
   1989   The Role of Wild Resources in Small-Scale Agricultural Systems: Tales from the Lakes and the Plains. In *Bad Year Economics,* edited by Paul Halstead and John O'Shea, pp. 57–67. Cambridge University Press, Cambridge.

O'Steen, Lisa D., Kristen J. Gremillion, and R. Jerald Ledbetter
   1991   *Archaeological Testing of Five Sites in the Big Sinking Creek Oil Field, Lee County, Kentucky.* Report submitted to Cecil R. Ison, Daniel Boone National Forest, Stanton, Kentucky.

Pearsall, Deborah M.
   1989   *Paleoethnobotany: A Handbook of Procedures.* Academic Press, San Diego.

Peebles, Christopher S.
   1974   *Moundville: The Organization of a Prehistoric Community and Culture.* Ph.D. dissertation, University of California, Santa Barbara. University Microfilms, Ann Arbor, Michigan.
   1983   Moundville: Late Prehistoric Sociopolitical Organization in the Southeastern United States. *American Ethnological Society Proceedings* 1979:183–98.
   1986   Paradise Lost, Strayed, and Stolen: Prehistoric Social Devolution in the Southeast. In *The Burden of Being Civilized,* edited by M. Richardson and M. Webb, pp. 24–40. Southern Anthropological Society Proceedings 18. University of Georgia Press, Athens.

1987a   Moundville from 1000 to 1500 A.D. as Seen from 1840 to
             1985 A.D. In *Chiefdoms in the Americas,* edited by R. Dren-
             nan and C. Uribe, pp. 21–42. University Press of Amer-
             ica, Lanham, Maryland.
1987b   The Rise and Fall of the Mississippian in Western Ala-
             bama: The Moundville and Summerville Phases, A.D.
             1000 to 1600. *Mississippi Archaeology* 22:1–31.

Perzigian, Anthony J., Patricia A. Tench, and Donna Braun
1984   Prehistoric Health in the Ohio River Valley. In *Paleo-
             pathology at the Origins of Agriculture,* edited by Mark
             Nathan Cohen and George J. Armelagos, pp. 347–66. Aca-
             demic Press, New York.

Petruso, Karl M., and Jere M. Wickens
1984   The Acorn in Aboriginal Subsistence in Eastern North
             America: A Report on Miscellaneous Experiments. In *Ex-
             periments and Observations on Aboriginal Wild Plant Food
             Utilization in Eastern North America,* edited by Patrick J.
             Munson, pp. 360–78. Indiana Historical Society Prehis-
             tory Research Series No. VI(2). Indianapolis.

Phillips, Paul C.
1961   *The Fur Trade.* 2 vols. University of Oklahoma Press,
             Norman.

Pillsbury, Richard
1983   The Europeanization of the Cherokee Settlement Land-
             scape Prior to Removal: A Georgia Case Study. *Geoscience
             and Man* 23:59–69.

Poesch, Jessie
1988   *The Art of the Old South: Painting, Sculpture, Architecture and
             the Products of Craftsmen, 1560–1860.* Harrison House,
             New York.

Popper, Virginia S.
1988   Selecting Quantitative Measurements in Paleoethnobot-
             any. In *Current Paleoethnobotany: Analytical Methods and Cul-
             tural Interpretations of Archaeological Plant Remains,* edited
             by C. A. Hastorf and V. S. Popper, pp. 53–71. University
             of Chicago Press, Chicago.

Popper, Virginia S., and Christine A. Hastorf
1988   Introduction. In *Current Paleoethnobotany: Analytical Meth-
             ods and Cultural Interpretations of Archaeological Plant Re-
             mains,* edited by C. A. Hastorf and V. S. Popper, pp. 1–16.
             University of Chicago Press, Chicago.

Porsild, A. Erling, and W. J. Cody

1979 *Vascular Plants of Continental Northwest Territorial Canada.* National Museum of Natural Sciences, National Museums of Canada, Ottawa.

Postgate, J. N., and M. A. Powell

1984 Preface. *Bulletin of Sumerian Agriculture* 1:v.

Potter, Stephen R., and Gregory A. Waselkov

1994 Whereby We Shall Enjoy Their Cultivated Places. In *Historic Archaeology of the Chesapeake,* edited by Paul A. Shackel and Barbara J. Little, pp. 23–33. Smithsonian Institution Press, Washington, D.C.

Pound, Merritt B.

1951 *Benjamin Hawkins: Indian Agent.* University of Georgia Press, Athens.

Powell, Mary Lucas

1988 *Status and Health in Prehistory: A Case Study of the Moundville Chiefdom.* Smithsonian Institution Press, Washington, D.C.

Prentice, Guy

1986 Origins of Plant Domestication in the Eastern United States: Promoting the Individual in Archaeological Theory. *Southeastern Archaeology* 5:103–19.

Pumpelly, R., editor

1908 *Explorations in Turkestan: Expedition of 1904—Prehistoric Civilizations of Anau: Origins, Growth, and Influence of Environment.* Carnegie Institution of Washington, Washington, D.C.

Quinn, David B., editor

1955 *The Roanoke Voyages.* Second Series No. 54. Hakluyt Society, London.

Radford, Albert E., Harry E. Ahles, and C. Ritchie Bell

1968 *Manual of the Vascular Flora of the Carolinas.* University of North Carolina Press, Chapel Hill.

Redding, Richard W.

1988 A General Explanation of Subsistence Change: From Hunting and Gathering to Food Production. *Journal of Anthropological Archaeology* 7:56–97.

Reice, Seth R.

1994 Nonequilibrium Determinants of Biological Community Structure. *American Scientist* 82:424–35.

Reidhead, Vann A.

1976 *Optimization and Food Procurement at the Prehistoric Leonard Haag Site, Indiana: A Linear Programming Approach.* Unpub-

References Cited

lished Ph.D. dissertation, Indiana University, Bloom-
ington.

Riley, Thomas J.

1987   Ridged Field Agriculture and the Mississippian Eco-
nomic Pattern. In *Emergent Horticultural Economies of the
Eastern Woodlands,* edited by W. F. Keegan, pp. 295–304.
Occasional Paper No. 7. Southern Illinois University,
Center for Archaeological Investigations, Carbondale.

Riley, Thomas J., Gregory R. Walz, Charles J. Bareis,
Andrew C. Fortier, and Kathryn E. Parker

1994   Accelerator Mass Spectrometry (AMS) Dates Confirm
Early *Zea mays* in the Mississippi River Valley. *American
Antiquity* 59:490–98.

Rindos, David

1984   *The Origins of Agriculture: An Evolutionary Perspective.* Academic Press, New York.

1989   Darwinism and its Role in the Explanation of Domestication. In *Foraging and Farming: The Evolution of Plant Exploitation,* edited by David R. Harris and Gordon C. Hillman, pp. 27–41. Unwin Hyman, London.

Rowland, Dunbar, and Albert G. Sanders, editors

1932   *Mississippi Provincial Archives, 1704–1743: French Dominion,*
vol. 3. Mississippi Department of Archives and History,
Jackson.

Russell, E.

1983   Indian-Set Fires in the Forests of the Northeastern
United States. *Ecology* 64(1):78–88.

Sahlins, Marshall

1972   *Stone Age Economics.* Aldine, Chicago.

Sahlins, Marshall, and E. R. Service, editors

1960   *Evolution and Culture.* University of Michigan Press, Ann
Arbor.

Scarry, C. Margaret

1986   *Change in Plant Procurement and Production during the Emergence of the Moundville Chiefdom.* Ph.D. dissertation, University of Michigan. University Microfilms, Ann Arbor.

1988   Variability in Mississippian Crop Production Strategies.
Paper presented at the 45th Annual Meeting of the
Southeastern Archaeological Conference, New Orleans.

1993   Plant Remains from the Big Sandy Farms Site (1Tu552),
Tuscaloosa County, Alabama. In *Big Sandy Farms: A Prehis-*

toric Agricultural Community Near Moundville, Black Warrior Valley Floodplain, Tuscaloosa County, Alabama by H. B. Ensor, pp. 205–33. Report of Investigations No. 68. Alabama Museum of Natural History, Division of Archaeology, Tuscaloosa.

Forthcoming   *Excavations on the Northwest Riverbank at Moundville: Investigations of a Moundville I Residential Area.* Report of Investigations No. 72. Alabama Museum of Natural History, Division of Archaeology, Tuscaloosa.

Schoener, Thomas W.

1974   Resource Partitioning in Ecological Communities. *Science* 185:27–39.

Schopmeyer, C. S.

1974   *Seeds of Woody Plants in the United States.* Agriculture Handbook No. 450. United States Department of Agriculture, Washington, D.C.

Schroedl, Gerald F., editor

1986   *Overhill Cherokee Archaeology at Chota-Tanasee.* Report of Investigations 38. Department of Anthropology, University of Tennessee, Knoxville.

Schwanitz, F.

1967   *The Origin of Cultivated Plants.* Harvard University Press, Cambridge.

Scott, Susan L.

1981   Economic and Organizational Aspects of Deer Procurement during the Late Prehistoric Period. Paper presented at the 38th Annual Southeastern Archaeological Conference, Asheville, North Carolina.

Segelquist, Charles A., and Walter E. Green

1968   Deer Food Yields in Four Ozark Forest Types. *Journal of Wildlife Management* 32:330–37.

Shipek, Florence

1989   An Example of Intensive Plant Husbandry: The Kumeyaay of Southern California. In *Foraging and Farming: The Evolution of Plant Exploitation,* edited by D. R. Harris and G. C. Hillman, pp. 159–67. Unwin Hyman, London.

Simmons, Linda Crocker

1983   The Emerging Nation, 1790 to 1830. In *Painting in the South: 1564–1980,* compiled by David S. Bundy, pp. 42–71. Virginian Museum of Fine Arts, Richmond.

Singer, Clay, and John E. Atwood

1987   *Phase II Archaeological Testing of a Portion of Site CA-SLO-165 in the City of Morro Bay, San Luis Obispo County, California.* Report submitted to the City of Morro Bay, California, November 15, 1987.

Slobodkin, Lawrence B.

1973   On the Inconstancy of Ecological Efficiency and the Form of Ecological Theories. In *Growth by Intussusception: Ecological Essays in Honor of G. Evelyn Hutchinson,* edited by E. S. Deevey, pp. 293–305. Transactions of the Connecticut Academy of Arts and Sciences, vol. 44. New Haven, Connecticut.

Smith, Bruce D.

1984   *Chenopodium* as a Prehistoric Domesticate in Eastern North America: Evidence from Russell Cave, Alabama. *Science* 226:165–67.

1985a   The Role of *Chenopodium* as a Domesticate in Pre-Maize Garden Systems of the Eastern United States. *Southeastern Archaeology* 4:51–72.

1985b   *Chenopodium berlandieri* ssp. *jonesianum:* Evidence for a Hopewellian Domesticate from Ash Cave, Ohio. *Southeastern Archaeology* 4:107–33.

1987a   The Economic Potential of *Chenopodium berlandieri* in Prehistoric Eastern North America. *Journal of Ethnobiology* 7:29–54.

1987b   Independent Domestication of Indigenous Seed-Bearing Plants in Eastern North America. In *Emergent Horticultural Economies of the Eastern Woodlands,* edited by William F. Keegan, pp. 3–47. Occasional Paper No. 7. Center for Archaeological Investigations, Southern Illinois University at Carbondale.

1989   Origins of Agriculture in Eastern North America. *Science* 246:1566–71.

1992a   The Floodplain Weed Theory of Plant Domestication in Eastern North America. In *Rivers of Change: Essays on Early Agriculture in Eastern North America,* by B. D. Smith, pp. 19–33. Smithsonian Institution Press, Washington, D.C.

1992b   Prehistoric Plant Husbandry in Eastern North America. In *The Origins of Agriculture: An International Perspective,* edited by C. W. Cowan and P. J. Watson, pp. 101–9. Smithsonian Institution Press, Washington, D.C.

1992c   *Rivers of Change, Essays on Early Agriculture in Eastern North America.* Smithsonian Institution Press, Washington, D.C.

Smith, Bruce D., and C. Wesley Cowan

1987   Domesticated *Chenopodium* in Prehistoric Eastern North America: New Accelerator Dates from Eastern Kentucky. *American Antiquity* 52:355–57.

Smith, B. D., C. W. Cowan, and M. P. Hoffman

1992   Is It an Indigene or a Foreigner? In *Rivers of Change: Essays on Early Agriculture in Eastern North America,* by B. D. Smith, pp. 67–100. Smithsonian Institution Press, Washington, D.C.

Smith, B. D., and V. A. Funk

1985   A Newly Described Subfossil Cultivar of *Chenopodium* (Chenopodiaceae). *Phytologia* 57:445–47.

Smith, David V.

1992   The Amino Acid Content of Seeds of Pre-Maize Crop Plants. Ms. on file, Department of Anthropology, National Museum of Natural History, Washington, D.C.

Smith, Eric A.

1991   Risk and Uncertainty in the "Original Affluent Society": Evolutionary Ecology of Resource-Sharing and Land Tenure. In *Hunters and Gatherers.* Vol. 1 of *History, Evolution and Social Change,* edited by Tim Ingold, David Riches, and James Woodburn, pp. 222–51. Berg, New York.

Smith, Eric A., and Bruce Winterhalder

1992   Natural Selection and Decision-Making: Some Fundamental Principles. In *Evolutionary Ecology and Human Behavior,* edited by E. A. Smith and B. Winterhalder, pp. 25–60. Aldine de Gruyter, Hawthorne, New York.

Smith, Eric A., and Bruce Winterhalder, editors

1992   *Evolutionary Ecology and Human Behavior.* Aldine de Gruyter, Hawthorne, New York.

Sork, Victoria L.

1983   Mast-fruiting in Hickories and Availability of Nuts. *American Midland Naturalist* 109:81–88.

Speth, John D., and Katherine A. Spielmann

1983   Energy Source, Protein Metabolism, and Hunter-Gatherer Subsistence Strategies. *Journal of Anthropological Archaeology* 2:1–31.

Stafford, C. Russell

1985   Campbell Hollow and Early-Middle Archaic Settlement Systems. In *The Campbell Hollow Archaic Occupations: A*

References Cited

Study of Intrasite Spatial Structure in the Lower Illinois Valley, edited by C. R. Stafford, pp. 236–54. Center for American Archaeology, Kampsville Archaeological Center, Kampsville, Illinois.

1991　Archaic Period Logistical Foraging Strategies in West-Central Illinois. *Midcontinental Journal of Archaeology* 16:212–45.

Stephens, D. W.

1990　Risk and Incomplete Information in Behavioral Ecology. In *Risk and Uncertainty in Tribal and Peasant Economies,* edited by Elizabeth Cashdan, pp. 19–46. Westview Press, Boulder, Colorado.

Stephens, D. W., and Eric L. Charnov

1982　Optimal Foraging: Some Simple Stochastic Models. *Behavioral Ecology and Sociobiology* 10:251–63.

Stephens, D. W., and John R. Krebs

1986　*Foraging Theory.* Princeton University Press, Princeton, N. J.

Steponaitis, Vincas P.

1983　*Ceramics, Chronology, and Community Patterns: An Archaeological Study at Moundville.* Academic Press, New York.

1991　Contrasting Patterns of Mississippian Development. In *Chiefdoms: Power, Economy, and Ideology,* edited by T. Earle, pp. 193–228. Cambridge University Press, Cambridge.

1992　Excavations at 1Tu50, an Early Mississippian Center near Moundville. *Southeastern Archaeology* 11:1–13.

Steward, Julian

1936　The Economic and Social Basis of Primitive Bands. In *Essays in Anthropology Presented to A. L. Kroeber,* edited by R. H. Lowie, pp. 331–50. University of California Press, Berkeley.

1955　*Theory of Culture Change.* University of Illinois Press, Urbana.

Stewart, Robert B., and William Robertson IV

1973　Application of the Flotation Technique in Arid Areas. *Economic Botany* 27:114–16.

Struever, Stuart

1968　Flotation Techniques for the Recovery of Small-Scale Archaeological Remains. *American Antiquity* 33:353–62.

Stuiver, M., and P. J. Reimer

1993　Extended [14]C Data Base and Revised CALIB 3.0 [14]C Age Calibration Program. *Radiocarbon* 35:215–30.

Styles, Bonnie W.
  1986   Aquatic Exploitation in the Lower Illinois River Valley:
                The Role of Paleoecologial Change. In *Foraging, Collect-
                ing, and Harvesting: Archaic Period Subsistence and Settlement in
                the Eastern Woodlands,* edited by Sarah W. Neusius, pp. 145–
                74. Center for Archaeological Investigations Occasional
                Paper No. 6. Southern Illinois University, Carbondale.

Swallen, J. R.
  1940   *Contributions Toward a Flora of Nevada, No. 1: Gramineae of
                Nevada.* Division of Plant Explorations and Introductions,
                United States Department of Agriculture, Washington, D.C.

Swetnam, Thomas W., and Julio L. Betancourt
  1990   Fire: Southern Oscillation Relations in the Southwestern
                United States. *Science* 249:1017–20.

Talalay, Laurie, Donald R. Keller, and Patrick J. Munson
  1984   Hickory Nuts, Walnuts, Butternuts, and Hazelnuts: Obser-
                vations and Experiments Relevant to Their Aboriginal
                Exploitation in Eastern North America. In *Experiments
                and Observations on Aboriginal Wild Plant Food Utilization in
                Eastern North America,* edited by Patrick J. Munson, pp.
                338–59. Indiana Historical Society Prehistory Research
                Series No. VI (2). Indianapolis.

Tapley, W. T., W. D. Enzie, and G. P. Eseltine
  1937   *The Cucurbits.* Vol. 1, part IV of *The Vegetables of New York.*
                Report of the New York State Agricultural Experiment
                Station, Albany.

Timbrook, Jan, John R. Johnson, and David D. Earle
  1982   Vegetation Burning by the Chumash. *Journal of California
                and Great Basin Anthropology* 4:162–86.

Tsukada, M., Shinya Sugita, and Yorko Tsukada
  1986   Oldest Primitive Agriculture and Vegetational Environ-
                ments in Japan. *Nature* 322:632–34.

Tukey, John W.
  1977   *Exploratory Data Analysis.* Addison-Wesley, Cambridge.

United States Department of Agriculture (USDA)
  1984   *Composition of Foods: Nut and Seed Products.* Agricultural
                Handbook No. 8–12. U.S. Department of Agriculture,
                Washington, D.C.

United States Department of Agriculture (USDA) Forest Service
  1948   *Woody-Plant Seed Manual.* Miscellaneous Publication No.
                654. U.S. Department of Agriculture Forest Service, Wash-
                ington, D.C.

Usner, Daniel H., Jr.
    1985   American Indians on the Cotton Frontier: Changing Economic Relations with Citizens and Slaves in the Mississippi Territory. *Journal of American History* 72:297–317.

van der Donck, Adriaen
    1846   *A Description of the New Netherlands* [reprinted 1968]. New York Historical Society Collection Series No. 2, vol. 1, pp. 125–242. Syracuse University Press, Syracuse.

van Zeist, W., K. Wsylikowa, and K.-E. Behre, editors
    1991   *Progress in Old World Paleoethnobotany.* A. A. Balkema, Rotterdam.

Velleman, Paul F., and David C. Hoaglin
    1981   *Applications, Basics, and Computing of Exploratory Data Analysis.* Duxbury Press, Boston.

Vogel, Joseph D., and Jean Allan
    1985   Mississippian Fortifications at Moundville. *Archaeology* 38(5):62–63.

Wagner, Gail E.
    1982   Testing Flotation Recovery Rates. *American Antiquity* 47:127–32.
    1986   The Corn and Cultivated Beans of the Fort Ancient Indians. *Missouri Archaeologist* 47:107–35.
    1987   *Uses of Plants by the Fort Ancient Indians.* Unpublished Ph.D. dissertation, Department of Anthropology, Washington University, St. Louis.
    1988   Comparability Among Recovery Techniques. In *Current Paleoethnobotany,* edited by C. Hastorf and V. Popper, pp. 17–35. University of Chicago Press, Chicago.

Ward, H. Trawick
    1965   Correlation of Mississippian Sites and Soil Types. *Southeastern Archaeological Conference Bulletin* 3:42–48.

Waselkov, Gregory A.
    1981   *Lower Tallapoosa River Cultural Resources Survey: Phase I Report.* Manuscript submitted to the Alabama Historical Commission, Montgomery.
    1988   Lamhatty's Map. *Southern Exposure* 16(2):23–29.
    1989   Indian Maps of the Colonial Southeast. In *Powhatan's Mantle: Indians in the Colonial Southeast,* edited by P. H. Wood, G. A. Waselkov, and M. T. Hatley, pp. 292–343. University of Nebraska Press, Lincoln.
    1992   English and American Trade with the Southeastern Indians, 1700–1800. Paper presented at the 13th Mid-

South Archaeological Conference, Moundville, Alabama.

Waselkov, Gregory A., John W. Cottier, and Craig T. Sheldon, Jr.

1990   *Archaeological Excavations at the Early Historic Creek Indian Town of Fusihatchee (Phase I, 1988–1989)*. Report prepared by Auburn University (Auburn), Auburn University (Montgomery), and the University of South Alabama for the National Science Foundation (Grant No. BNS-8718934).

Watson, Patty Jo

1969   *The Prehistory of Salts Cave, Kentucky.* Reports of Investigations No. 16. Illinois State Museum, Springfield.

1976   In Pursuit of Prehistoric Subsistence: A Comparative Account of Some Contemporary Flotation Techniques. *Midcontinental Journal of Archaeology* 1:77–100.

1985   The Impact of Early Horticulture in the Upland Drainages of the Midwest and the Midsouth. In *Prehistoric Food Production in North America,* edited by R. Ford, pp. 99–147. Anthropological Papers No. 75. Museum of Anthropology, University of Michigan, Ann Arbor.

1989   Early Plant Cultivation in the Eastern Woodlands of North America. In *Foraging and Farming: The Evolution of Plant Exploitation,* edited by D. Harris and G. Hillman, pp. 555–71. Unwin-Hyman, London.

1991a   Early Plant Cultivation in the Southwest and in the Eastern Woodlands: Pattern and Process. In *Puebloan Past and Present: Papers in Honor of Steward Peckham,* edited by M. Duran and D. Kirkpatrick, pp. 183–200. The Archaeological Society of New Mexico, Albuquerque.

1991b   Origins of Food Production in Western Asia and Eastern North America: A Consideration of Interdisciplinary Research in Anthropology and Archaeology. In *Quaternary Landscapes,* edited by L. Shane and E. Cushing, pp. 1–37. University of Minnesota Press, Minneapolis.

Watson, Patty Jo, editor

1974   *Archeology of the Mammoth Cave Area.* Academic Press, New York.

Watson, Patty Jo, and Mary Kennedy

1991   The Development of Horticulture in the Eastern Woodlands of North America: Women's Role. In *Engendering Archaeology: Women and Prehistory,* edited by J. Gero and M. Conkey, pp. 255–75. Basil Blackwell, Oxford.

Watt, B. K., and D. A. Merrill

    1975   *Handbook of the Nutritional Contents of Foods.* Dover, New York.

Weaver, Martin

    1971   A New Water Separation Process for Soil from Archaeological Excavation. *Anatolian Studies* 21:65–68.

Webb, William S., and William D. Funkhouser

    1936   Rock Shelters in Menifee County, Kentucky. *Reports in Anthropology and Archaeology* 3:105–67. University of Kentucky, Lexington.

Weber, W. A., and R. C. Willmann

    1992   *Catalog of the Colorado Flora: A Biodiversity Baseline.* University Press of Colorado, Boulder.

Welch, Paul D.

    1989   Chronological Markers and Imported Items from the Roadway Excavations at Moundville. Paper presented at the 46th Annual Southeastern Archaeological Conference, Tampa.

    1990   Mississippian Emergence in West Central Alabama. In *Mississippian Emergence,* edited by B. Smith, pp. 197–225. Smithsonian Institution Press, Washington, D.C.

    1991   *Moundville's Economy.* University of Alabama Press, Tuscaloosa.

Welsh, Stanley L., N. Duane Atwood,
Sherel Goodrich, and Larry C. Higgins, editors

    1987   *A Utah Flora.* Great Basin Naturalists Memoirs No. 9. Brigham Young University, Provo, Utah.

White, Leslie A.

    1949   *The Science of Culture.* Grove Press, New York.

    1959   *The Evolution of Culture.* McGraw-Hill, New York.

White, Richard

    1983   *The Roots of Dependency: Subsistence, Environment, and Social Change Among the Choctaws, Pawnees, and Navajos.* University of Nebraska Press, Lincoln.

Wilde, James D., and Deborah E. Newman

    1989   Late Archaic Corn in the Eastern Great Basin. *American Antiquity* 91:712–17.

Wilkinson, Leland

    1990   *SYGRAPH: The System for Graphics.* SYSTAT, Evanston, Illinois.

Williams, Samuel C., editor

    1930   *Adair's History of the American Indians.* Promontory Press, New York.

Wills, W. H.
  1988a   *Early Prehistoric Agriculture in the American Southwest.*
             School of American Research, Santa Fe.
  1988b   Early Agriculture and Sedentism in the American South-
             west: Evidence and Interpretations. *Journal of World Prehis-
             tory* 2:445–88.
Wilms, Douglas C.
  1974   Cherokee Settlement Patterns in Nineteenth Century
             Georgia. *Southeastern Geographer* 14:46–53.
Wilson, Gilbert L.
  1987   *Buffalo Bird Woman's Garden: Agriculture of the Hidatsa Indians*
             [orig. 1917]. Minnesota Historical Society Press, St. Paul.
Wilson, H. D.
  1981   Domesticated *Chenopodium* of the Ozark Bluff Dweller.
             Economic Botany 35:233–39.
Wilson, R.
  n.d.   Cave Research Foundation Squash/Gourd Inventory. Un-
             published worksheets in possession of the author.
Winter, Joseph C.
  1973   The Distribution and Development of Fremont Maize
             Agriculture: Some Preliminary Interpretations. *American
             Antiquity* 38(4):439–52.
  1974   *Aboriginal Agriculture in the Southwest and Great Basin.*
             Ph.D. dissertation, Department of Anthropology, Univer-
             sity of Utah, Salt Lake City.
  1976   The Processes of Farming Diffusion in the Southwest
             and Great Basin. *American Antiquity* 41(4):412–29.
  1982   Alternative Views of the "Diffusion" of Eight-Row
             Maize: A Study of Corn From Three Basketmaker Sites
             near Bluff, Utah. In *Basketmaker Settlement and Subsistence
             Along the San Juan River, Utah: The U.S. 163 Archaeological
             Project,* edited by Robert B. Neily et al. Antiquities Sec-
             tion, Utah Division of State History, Salt Lake City.
  1991   Prehistoric and Historic Native American Tobacco Use:
             An Overview. Paper presented at the Society for Ameri-
             can Archaeology Annual Meeting, New Orleans.
Winter, Joseph C., and Patrick Hogan
  1986   Plant Husbandry in the Great Basin and Adjacent North-
             ern Colorado Plateau. *University of Utah Anthropological Pa-
             pers* 110:117–44.
Winterhalder, Bruce
  1980   Environmental Analysis in Human Evolution and Adapta-
             tion Research. *Human Ecology* 8:135–70.

References Cited

1986   Diet Choice, Risk, and Food Sharing in a Stochastic Environment. *Journal of Anthropological Archaeology* 5:369–92.

1990   Open Field, Common Pot: Harvest Variability and Risk Avoidance in Agricultural and Foraging Societies. In *Risk and Uncertainty in Tribal and Peasant Economies,* edited by Elizabeth Cashdan, pp. 67–87. Westview Press, Boulder, Colorado.

1993   Work, Resource and Population in Foraging Societies. *Man* (n.s.) 28:321–40.

Winterhalder, B., W. Baillargeon, F. Cappalleto, R. Daniel, and C. Prescott

1988   The Population Dynamics of Hunter-Gatherers and their Prey. *Journal of Anthropological Archaeology* 7:289–328.

Winterhalder, Bruce, and Carol Goland

1993   On Population, Foraging Efficiency, and Plant Domestication. *Current Anthropology* 34:710–15.

Winterhalder, Bruce, and Eric A. Smith

1992   Evolutionary Ecology and the Social Sciences. In *Evolutionary Ecology and Human Behavior,* edited by Eric A. Smith and Bruce Winterhalder, pp. 3–23. Aldine de Gruyter, Hawthorne, New York.

Woodburn, James

1982   Egalitarian Societies. *Man* (n.s.) 17:431–51.

Woods, William I.

1987   Maize Agriculture and the Late Prehistoric: A Characterization of Settlement Location Strategies. In *Emergent Horticultural Economies of the Eastern Woodlands,* edited by W. F. Keegan, pp. 275–94. Occasional Paper No. 7. Center for Archaeological Investigations, Southern Illinois University, Carbondale.

Wright, J. Leitch

1967   *William Augustus Bowles: Director General of the Creek Nation.* University of Georgia Press, Athens.

1986   *Creeks and Seminoles: The Destruction of the Muscogulge People.* University of Nebraska Press, Lincoln.

Wymer, Dee Anne

1990   Archaeobotany. In *Childers and Woods: Two Late Woodland Sites in the Upper Ohio Valley, Mason County, Kentucky,* edited by Michael J. Schott, pp. 487–616. University of Kentucky Program for Cultural Resource Assessment Archaeological Report No. 200. Lexington.

1993   Cultural Change and Subsistence: The Middle Woodland and Late Woodland Transition in the Mid-Ohio Valley. In

Foraging and Farming in the Eastern Woodlands, edited by
C. Margaret Scarry, pp. 138–56. University Press of Florida, Gainesville.

Yabuno, Tomosaburo

1966   Biosystematic Study of the Genus *Echinochloa. Japanese
Journal of Botany* 19:277–323.

1987   Japanese Barnyard Millet (*Echinochloa utilis,* Poaceae) in Japan. *Economic Botany* 41:484–93.

Yarnell, Richard A.

1963   Reciprocity in Cultural Ecology. *Economic Botany*
17:333–36.

1964   *Aboriginal Relationships Between Culture and Plant Life in
the Upper Great Lakes Region.* Anthropological Papers No.
23. Museum of Anthropology, University of Michigan,
Ann Arbor.

1965   Implications of Distinctive Flora on Pueblo Ruins. *American Anthropologist* 67:662–74.

1969   Contents of Human Paleofeces. In *The Prehistory of Salts
Cave, Kentucky,* by Patty Jo Watson, pp. 41–54. Reports of
Investigations No. 16. Illinois State Museum, Springfield.

1970   Paleo-Ethnobotany in America. In *Science in Archaeology,*
edited by D. Brothwell and E. Higgs, pp. 215–28. Praeger,
New York.

1972   *Iva annua* var. *macrocarpa:* Extinct American Cultigen?
American Anthropologist 74:335–41.

1974a   Intestinal Contents of the Salts Cave Mummy and Analysis of the Initial Salts Cave Flotation Series. In *Archaeology
of the Mammoth Cave Area,* edited by Patty Jo Watson, pp.
109–12. Academic Press, New York.

1974b   Plant Foods and Cultivation of the Salts Cavers. In *Archaeology of the Mammoth Cave Area,* edited by Patty Jo
Watson, pp. 113–22. Academic Press, New York.

1978   Domestication of Sunflower and Sumpweed in Eastern
North America. In *The Nature and Status of Ethnobotany,*
edited by Richard I. Ford, pp. 289–300. Anthropological
Papers No. 67. University of Michigan Museum of Anthropology, Ann Arbor.

1981   Inferred Dating of Ozark Bluff Dweller Occupations
Based on Achene Size of Sunflower and Sumpweed. *Journal of Ethnobiology* 1:55–60.

1986   A Survey of Prehistoric Crop Plants in Eastern North
America. In "New World Paleoethnobotany: Collected

Papers in Honor of Leonard W. Blake." *The Missouri Archaeologist* 47:47–59.

1993   The Importance of Native Crops During the Late Archaic and Woodland. In *Foraging and Farming in the Eastern Woodlands,* edited by C. M. Scarry, pp. 13–26. University of Florida Press, Gainesville.

1994   Investigations Relevant to the Native Development of Plant Husbandry in Eastern North America: A Brief and Reasonably True Account. In *Agricultural Origins and Development in the Midcontinent,* edited by William Green, pp. 7–24. Report No. 19. Office of the State Archaeologist, University of Iowa, Iowa City.

In Press   *Sunflower, Sumpweed, Small Grains and Crops of Lesser Status.* Written for the Smithsonian Handbook of North American Indians.

Yarnell, Richard A., and M. Jean Black
1985   Temporal Trends Indicated by a Survey of Archaic and Woodland Plant Food Remains from Southeastern North America. *Southeastern Archaeology* 4:93–107.

Yasuda, Y.
1978   *Prehistoric Environment in Japan: Palynological Approach.* Institute of Geography, Faculty of Science, Tohoku University, Sendai, Japan.

Yesner, David R.
1981   Archaeological Applications of Optimal Foraging Theory: Harvest Strategies of Aleut Hunter-Gatherers. In *Hunter-Gatherer Foraging Strategies: Ethnographic and Archeological Analyses,* edited by Bruce Winterhalder and Eric Alden Smith, pp. 148–70. University of Chicago Press, Chicago.

Zawacki, April A., and Glenn Hausfater
1969   *Early Vegetation of the Lower Illinois Valley.* Illinois State Museum Reports of Investigations No. 17. Springfield.

Zitomersky, Joseph
1992   The Form and Function of French–Native American Relations in Early Eighteenth-Century French Colonial Louisiana. In *Proceedings of the Fifteenth Meeting of the French Colonial Historical Society,* edited by Patricia Galloway and Philip P. Boucher, pp. 154–77. University Press of America, Lanham, Maryland.

CONTRIBUTORS

C. Wesley Cowan is Curator of Archaeology at the Cincinnati Museum of Natural History. He has long been interested in the evolution of agricultural economies in eastern North America. His research centers on early horticulturalists in the Cumberland Plateau of eastern Kentucky and the maize-based Fort Ancient cultures of the central Ohio Valley.

Gary W. Crawford is Professor of Anthropology at the University of Toronto. He has published a monograph and numerous articles on the prehistory and paleoethnobotany of Japan. He has also investigated and written about issues on the prehistory of eastern North America. In recent years his research has focused on the archaeology of the immediate predecessors of the Ainu of northeastern Japan. His current research, supported by the Social Sciences and Humanities Research Council of Canada and Earthwatch, addresses the beginning of agriculture in southern Ontario, Canada.

Gayle J. Fritz is Assistant Professor of Anthropology at Washington University in St. Louis and holds a joint appointment in Biology. Her work explores prehistoric human-plant interrelationships, especially the cultural, biological, and ecological aspects of subsistence continuity and change. A focus is on the processes and sequences that led to the development of agricultural systems in eastern North America. Currently, she works in the central and lower regions of the Mississippi Valley and in the Trans-Mississippi South, modeling the transition to farming and subsequent intensification of maize-based agricultural systems.

Paul S. Gardner received his doctorate in anthropology from the University of North Carolina at Chapel Hill, where he specialized in archaeobotanical and evolutionary ecological studies of Eastern Woodlands subsistence change. As an archaeologist he has directed field research in Virginia and North Carolina and as a consulting archaeobotanist has analyzed plant remains from sites in seven states. Currently he is Adjunct Assistant Professor of Anthropology at The Ohio State University and is Midwest Regional Director for The Archaeological Conservancy, the nation's private, not-for-profit organization dedicated to preserving archaeological sites through permanent acquisition.

Carol Goland is Affiliated Scholar and Visiting Assistant Professor at Denison University, where she teaches courses in the Sociology/Anthropology Department and the Environmental Studies Program. Her recent work focuses on the human ecology of agricultural production in the peasant communities of Cuyo Cuyo in the Peruvian Andes. Aspects of this work are summarized in "Field Scattering as Agricultural Risk Management: A Case Study from Cuyo Cuyo, Department of Puno, Peru" (*Mountain Research and Development,* 1993) and "Agricultural Risk Management Through Diversity: Field Scattering in Cuyo Cuyo, Peru" (*Culture and Agriculture,* 1993). Her interest in forager subsistence strategies grew out of earlier work on the archaeology of the eastern United States Archaic.

Kristen J. Gremillion is Assistant Professor of Anthropology at The Ohio State University. She received her doctorate from the University of North Carolina at Chapel Hill, in 1989. Her research specialties include paleoethnobotany, prehistoric diet and subsistence, and the origins of agriculture. She is currently investigating the development of premaize farming systems in the uplands of eastern Kentucky and is involved in the multidisciplinary study of human paleofeces as a source of information about the diet, health, and genetics of ancient populations.

Julia E. Hammett is Research Scientist with Muwekma Ohlone Tribe in San Jose, California, and Research Associate with Stanford Campus Archaeology Program. She has conducted paleoethnobotanical and landscape analyses in four regions of North America: the Southwest, California, the Southeast, and the Great Basin. Her research combines ecological, archaeological, and historical data to investigate prehistoric and historic landscapes and traditional land use patterns. She currently conducts research in the San Francisco Bay area.

C. Margaret Scarry is Assistant Professor of Anthropology and Research Associate of the Research Laboratories of Anthropology at the University of North Carolina. She received her doctorate in anthropology in 1986 from the University of Michigan. Her research focuses on the foodways of the late prehistoric and early historic peoples of southeastern North America. She has published articles on investigation of archaeological plant remains from the Moundville polity in Alabama and from several Spanish colonial sites in Florida. In addition, she has analyzed plant remains from the Parkin site in Arkansas and from sites in Indiana, Kentucky, Maryland, Mississippi, Missouri, New York, South Carolina, and Virginia.

Bruce D. Smith is Senior Scientist and Director of the Archaeobiology Program, Department of Anthropology, National Museum of Natural History, Smithsonian Institution, Washington, D.C. His recent books include *The Mississippian Emergence* (1990), *Rivers of Change: Essays on the Early Agriculture of Eastern North America* (1992), and *The Emergence of Agriculture* (1994).

Vincas P. Steponaitis is Professor of Anthropology and Director of the Research Laboratories of Anthropology at the University of North Carolina at Chapel Hill. He is a specialist on the ancient history and archaeology of the American South and has written extensively about Moundville and other sites in this region.

Gregory A. Waselkov is Associate Professor of Anthropology at the University of South Alabama, where he is also Director of the Center for Archaeological Studies. He has written extensively on the Creek Indians of central Alabama during the colonial era and on French colonization along the Gulf Coast, and he has just co-edited a book on the Indian writings of William Bartram. Currently he is directing excavations at Old Mobile, site of a French town between 1702 and 1711, and is collaborating on a long-term study of oyster shell growth that could help determine seasonality of archaeological site occupations.

Patty Jo Watson is Edward Mallinckrodt Distinguished University Professor of Anthropology at Washington University in St. Louis. She has published several articles and books on cave and shell mound archaeology in west central Kentucky and continues to do field research focused on early agriculture in the Mammoth Cave area.

Bruce Winterhalder is Professor and Chairman, Department of Anthropology, and a member of the Ecology Curriculum, University of North Carolina at Chapel Hill. He has conducted fieldwork and published on the foraging strategies of boreal forest Cree living in northern

Ontario and has done research on agropastoral ecology among Quechua peasants living in the Peruvian Andes. He currently is investigating exchange behavior in hunter-gatherers and analyzing data on time allocation of Peruvian farmers. With Eric Smith, he is co-editor of *Hunter-Gatherer Foraging Strategies* (1981) and *Evolutionary Ecology and Human Behavior* (1992).

INDEX

Accelerator mass spectrometry (AMS): of chenopod from Newt Kash, 29, 73; of *Cucurbita* from Cloudsplitter, 68, 69, 81; of Middle Holocene cucurbits, 28; radiocarbon dating, xvi, 19, 20; of rice from Kazahari, 101; of seeds from Marble Bluff, 46; of sunflower from Hayes site, 55–56

Aché, 162

Acorn (*Quercus*): meal, 116, 165; in Moundville economy, 112, 115; nutritional composition of, 162; in prehistoric Japan, 94, 95; processing costs of, 165; used by Carolina Algonquians, 161; yield of, 163–64. *See also* Nut resources

Adaptation, xvi, 3, 124, 134; to European invasion of North America, 179, 191. *See also* Selection

Adaptationism. *See* Functionalist-adaptationist explanation

Agave (*Agave*), 21

Agnew shelter (Arkansas), 52

Agricultural fields: archaeological evidence for, 179; communal, 120, 180, 183, 187; dispersion of, 120, 123–24, 137–39, 149–50; historical records of, 179; location of, 179–84, 187–88; paired with household gardens, 180, 183, 187, 197; rotation of, 121; in southwestern United States, 199

Agricultural origins: 20, 125, 126, 157; computer simulation of, 152–53; in eastern North America, 1, 25–26, 124, 143, 177; explanation of, 2, 3; models of, 123, 127, 152–56; in Near East, 14–16, 21, 87; role of diffusion in, 27–28

Agriculture, 17, 124; defined, 9; disturbance associated with, 87, 88; evolution of, 120; maize-based, 115, 146, 149–50; nonindustrial, 137, 158; in prehistoric eastern North America, 1–2, 19, 63, 143–52; in prehistoric Japan, 86, 98–99; in prehistoric Near East, 14–16; in prehistoric southwestern United States, 20; monocrop, 87; rice, 88–89, 98–99; temperate-zone and tropical compared, 157.

See also Farming; Food production; Plant husbandry

Ainu, 88, 97, 101

Algonquians, 161

Allozyme analysis, 53

Alred shelter (Arkansas), 52

Amaranth (*Amaranthus* sp.), 21

Amur corktree (*Phellodendron amurense*), 93

Anadromous fish, 150

Anderson, Edgar, 144

Anthropogenesis, xvi; concept of, 86–88; evidence for, 89–90, 102–3; and grasses, 96; during Jomon period, 93–95; and landscape structure, 197; and textile plants, 200; and weeds, 177, 197, 200. *See also* Disturbance

Aomori (Japan), 89, 101

Aquatic resources, 59, 144, 145, 177

Archaeobotany, 4, 13; defined, xvi

Asch, David, 25

Asch, Nancy, 25

Ayni, 142

Barley (*Hordeum vulgare*), 95

Barnyard grass (*Echinochloa crus-galli*), 93, 103; cultivation of, 97–100

Barnyard millet (*Echinochloa utilis*), 103; carbonized mass of, 100; cultivation of, 97–100

Bartram, William, 201–2

Bat Cave (New Mexico), 19, 211

Beans (*Phaseolus vulgaris*), 195, 209; diffusion of, 211; ethnohistoric records of, 161

Beidha (Jordan), 15

Blackberry (*Rubus*), 120

Black Warrior Valley (Alabama), 107, 112

Blake, Leonard, 17, 18

Botanical Gardens site (Japan), 96

Bottle gourd (*Lagenaria siceraria*), 20, 21, 143; from Windover site, 54

Boxplots, 113

Braidwood, Robert J., 14, 18, 21

Broomcorn millet (*Panicum miliaceum*), 93, 98, 101, 103

Buckwheat (*Fagopyrum esculentum*), 93